Your Veterinary Wellness Journey

The resources offered in this workbook are an evidence-based toolbox that will equip veterinary professionals, students, and teams to meet the challenges of caregiving in a practical, relevant, and hopeful way.

Your Veterinary Wellness Journey builds on the essential work begun in the author's first book, *Creating Wellbeing and Building Resilience in the Veterinary Profession*, which introduced the care-giving challenges that may impact wellbeing, especially those unique to veterinary environments. This practical sequel capitalizes on collective, compassionate wisdom drawn from a global perspective on what tools, practices, and approaches will inform, prepare, and inspire all members of a veterinary team to be their healthiest selves, both at work and in their personal lives.

The workbook shares the intention of the first book of building efficacy, community, and creative collaboration, supporting each of us to be actively involved in contributing to the positive changes we wish to see in and around the veterinary profession as we meet the needs of animals and humans worldwide. It is an important resource for anyone working in the veterinary profession and general practice, as well as being a great book for post-graduates and undergraduates preparing for their career.

Your Veterinary Wellness Journey

An Interactive Guide to Thriving with Growth, Resilience, and Purpose

Sonja A. Olson

CRC Press
Taylor & Francis Group
Boca Raton London New York

CRC Press is an imprint of the
Taylor & Francis Group, an **informa** business

Designed cover image: Sonja A. Olson's original

First edition published 2027
by CRC Press
2385 NW Executive Center Drive, Suite 320, Boca Raton FL 33431

and by CRC Press
4 Park Square, Milton Park, Abingdon, Oxon, OX14 4RN

CRC Press is an imprint of Taylor & Francis Group, LLC

ISBN: 978-1-032-75717-9 (hbk)
ISBN: 978-1-032-75392-8 (pbk)
ISBN: 978-1-003-47536-1 (ebk)

DOI: 10.1201/9781003475361

Typeset in Palatino
by Apex CoVantage, LLC

SUPPORT MATERIAL https://www.routledge.com/9781032753928

Contents

Foreword

For more than three decades, a dedicated community of researchers, educators, and clinicians has worked to understand and address the impact of secondary traumatic stress and chronic strain on helping professionals. Across disciplines – health care, justice, education, social services – the same patterns emerge: dedicated individuals give deeply of themselves in this meaningful and demanding work, often at great personal cost. Veterinary professionals are no exception.

When I first had the opportunity to meet and talk with Dr. Sonja A. Olson, first as a participant in a course I was teaching and later as a colleague in a multidisciplinary think tank, I was immediately struck by her clarity of thought and her deep respect for those who care for others. She brings the analytical mind of a clinician together with the compassion of someone who understands, from experience, what it means to sustain oneself in complex and high-stakes work.

Dr. Olson's earlier book opened important conversations about wellbeing and compassion in veterinary medicine. This new volume builds on that foundation. It is evidence-informed, realistic, and infused with hope. She offers practical guidance for sustaining a meaningful career, acknowledging both the systemic pressures that shape the work and the small, daily actions that can restore steadiness and purpose.

Within our shared field of secondary traumatic stress, we have learned that wellbeing cannot rest on individual effort alone. Sustainable practice requires environments that support recovery, connection, and professional integrity. Dr. Olson's approach reflects this evolution. She brings the science of human sustainability to life, offering tools that are both accessible and aligned with the realities of veterinary teams today.

Your Veterinary Wellness Journey: An Interactive Guide to Thriving with Growth, Resilience, and Purpose is a companion for those navigating the highs and lows of veterinary life. It is a reminder that the goal is not perfection but steadiness, connection, and meaning. Dr. Olson offers a wise and compassionate path forward, one that strengthens not only individual practitioners but also the culture of care that sustains them. I hope this book offers a moment to pause, reflect, and carry on with renewed purpose in this demanding and meaningful work.

Françoise Mathieu, M.Ed., RP
Kingston, Ontario

About the Author

Dr. Sonja A. Olson (C-MMT, C-Neuroinformed Coach, MentorVet Certified Veterinary Mentor) is a credentialed veterinary wellness coach, educator, and emergency room veterinarian with nearly 30 years of experience. She specializes in helping veterinary individuals, teams, and leaders navigate personal and professional stress with more skill, optimize their holistic health, and cultivate emotional wellbeing. Combining evidence-based neuroplasticity coaching, mindfulness, and her clinical expertise, Dr. Olson creates compassionate, transformative spaces that empower her clients and audiences to thrive in their lives. She is an author of two books and a contributing author to many other veterinary textbooks and journal articles.

Through her tailored sessions, keynote presentations, webinars, retreats, and resilience-focused writings, she shares practical tools to manage change, enhance self-awareness, and foster healthy veterinary teams and culture. As a certified mindfulness meditation teacher, neuroinformed and trauma-informed coach, and Mental Health First Aid instructor, she is dedicated to inspiring calmer, kinder, and more resilient communities. Dr. Olson's mission is to harness the innate human potential for healing and growth, guiding veterinary professionals toward a healthier, optimized way of living.

Introduction: A Journey Toward Sustainable Wellbeing in Veterinary Medicine

Veterinary professionals are compassionate, intelligent, and adaptable. We solve complex medical puzzles, navigate unpredictable challenges, and give so much of ourselves to our patients, clients, and teams.

Yet when it comes to our own wellbeing, we often struggle to apply that same problem-solving mindset. I've lived the chaos of ER practice. I've stayed late to support a colleague in tears. I've felt both the incredible privilege and the weighty responsibility of being a veterinary caregiver. I know how easy it is to lose yourself in service to others. And I know how vital it is that we learn new ways forward – not just for ourselves but for the entire ecosystem of veterinary care.

What I have experienced and witnessed in other veterinary professionals moves my heart. The struggles to meet the most basic human needs for adequate sleep/rest, safety (physical and psychological), and meeting the body's needs for nutrition and movement are often ignored. Giving ourselves permission to "be" in our lives more than just exist in our veterinary "doing" can feel at odds with what we have come to understand it means to be a "successful," "committed" veterinary teammate and care provider. The real and perceived barriers to having an integrated life where we are living into our values, our purpose, and our version of a meaningful existence can feel daunting.

What gets in our way?

- **Not knowing where to start.** When life and work feel overwhelming, even small changes can seem impossible.
- **Guilt.** Setting boundaries around self-care can feel selfish, especially in a profession built on service and sacrifice.
- **Time scarcity.** The pressure to always do more, faster, makes prioritizing wellbeing feel unrealistic.
- **Exhaustion.** Deep, cellular-level fatigue leaves many feeling like they simply don't have the energy to make a change in themselves or in their lives.
- **Feeling alone on the journey.**

These challenges are real. But so are the solutions.

What if you had evidence-based strategies to increase your energy, efficiency, and cognitive function – both personally and professionally? What if you could build resilience in ways that feel practical and sustainable, rather than like just another thing on your to-do list?

I am here to help you do exactly that.

As an emergency clinician and ER team leader for over 20 years, I've lived through the stress, the sleepless nights, the emotional toll, and the deep sense of purpose that comes with this work. I've also seen firsthand and have personally experienced how the historical culture of veterinary medicine – its expectations, norms, and systemic pressures – can contribute to burnout and distress. Importantly, there are also many moments when the focused intention and energy of a veterinary team coming together to problem solve complex challenges in medicine and in culture have uplifted my energy and gave me genuine pride in our capabilities, particularly when we work together.

Veterinary medicine is a team sport. Also, regardless of where you consider yourself on the introvert-ambivert-extrovert scale, all humans are social animals. We are smarter and more capable when we come together. When we combine intellectual and physical strengths, we have more than the sum of the parts. Rather, our collective input allows us to take on, to push through, and to overcome. There is also a remembering of the joy in what we do when we feel that we are not alone on the journey of learning and persevering in life and in our chosen work.

This is the ethos that I wish to bring forward in this book. Imagine that we are embarking upon a challenging trek in an unpredictable, exciting, and beautiful landscape together. Before setting off on such an adventure, we would have taken the time to prepare: What tools and resources will be needed? What "nutrition" will support the sustained effort? Who on our team has what knowledge and skills to fortify the collective whole while on the trail? What will we do when the inevitable injury or obstacle (real or perceived) occurs? These are the considerations that make the difference between an exhausting slog versus a life-changing growth experience.

I am a realistic optimist. I have the belief that with preparedness, grit, and intelligent humility, I can meet many of life's challenges with a mindset of self-confidence and readiness for growth. I equally believe in the capabilities and capacities of my fellow veterinary colleagues around the world. Many of us share the values of service, compassion, and learning. We are curious and love a good challenge, particularly when we feel prepared and when we have a supportive community.

There are socio-economic and cultural challenges in the world today that have contributed to and perhaps intensified the existing stressors in veterinary medicine. Morale, worker engagement, client interactions, and even patient care are all affected. While systemic change is necessary, individual and team-level strategies remain vital. We need practical, actionable ways to build resilience and reclaim our sense of fulfillment in this profession.

That's where this workbook comes in.

This is not about quick fixes or empty wellness rhetoric. It's about equipping you with the tools, knowledge, and support to create a sustainable, fulfilling career in veterinary medicine – one step at a time.

Through this journey, we will explore:

- **Practical strategies** for managing stress, energy, and time
- **Mindset shifts** that make wellbeing a natural part of your professional identity
- **Tools** for fostering a supportive, psychologically safe workplace culture
- **Ways to reconnect** with the meaning and joy in your work

The content of this workbook is meant to be for all veterinary professionals and leaders regardless of the segment of veterinary medicine or role you work in at present. This workbook will provide language leveling, heightened awareness of topics and approaches, and a deepening of concepts that may already be familiar to you.

I invite you to lean into the "I don't know what I don't know" space as well as "I don't know this . . . YET!" Choose your adventure in this workbook by jumping onto the "learning" trail where you feel called to at this point in your own journey. You can go backward and forward, grab hold of content that feels supportive for you on this particular leg of the life trek. Be open to the discomfort and awkwardness that is part of growth. Again, you are strengthening your "muscles" for the endurance and heart-centered fortitude needed to follow the zigs and zags, ups and downs of the trail ahead.

You are not alone in this. Together, let's move toward a healthier and more sustainable way of being – both as veterinary professionals and as human beings.

Now, eat that energy bar, lace up those hiking boots, get that pack on your back, and let the shared adventure begin!

The Wellness Trek: The Path Towards Healthy Selves, Community, and Profession.

Glossary of Terms

Active listening: A communication skill that involves fully concentrating on, understanding, and responding to what someone is saying. This involves paying attention to both verbal and nonverbal cues and processing the meaning and intent behind the spoken words.

Anxiety disorder: An emotional state characterized by disproportionate feelings of tension, recurring intrusive thoughts or concerns, and physical changes like increased blood pressure or heart rate.

Boundaries:

- ***Outward-facing boundaries*:** Using your purpose, values, and needs to guide how you use your resources (e.g., time, money, energy).
- ***Inward-facing boundaries*:** Knowing what is and is not your responsibility to protect your sense of self (self-regard, self-worth, beliefs).

Boredom boreout syndrome: A psychological disorder that causes physical illness and low kinetic energy state (languishing) associated with low work-related arousal and mental underload.

Burnout: As defined as a workplace syndrome, physical and emotional exhaustion secondary to chronic occupational distress that includes excessive responsibilities or time at work.

- ***Frenetic subtype*** – typical of work overload and workers who are highly dedicated tending to work until exhaustion.
- ***Worn-out subtype*** – characterized by feelings of hopelessness, lack of control over work results or conditions, and/or lack of recognition/appreciation for their efforts.

Cognitive reframing: A technique used to shift one's mindset to look at a situation from a different perspective.

Collaborative leadership: A management style where leaders work alongside employees at all levels to achieve common goals. It emphasizes teamwork, cooperation, communication, and inclusivity.

Collective intelligence: A group's combined capacity and capability to perform a wide range and variety of tasks and solve diverse problems. Individual intelligence is extended to groups, highlighting the power of collaboration and shared knowledge, and is essential in consensus decision-making.

Compassion: Literally it means "to suffer together." It is a positive emotion that arises when confronted with another's suffering, followed by the motivational feeling to relieve that suffering.

Compassion fatigue: Chronic stress or exhaustion due to the physical, emotional, and mental strain of caregiving.

Compassion satisfaction: The positive consequences of helping behavior. The intrinsic sense of fulfillment derived from helping others and the positive feelings for colleagues and oneself resulting from the ability to assist others and to contribute.

Conflict avoidance: A method of reacting to conflict that attempts to avoid directly confronting the issue (*Note: this style of conflict management is a type of people-pleasing behavior*).

Constructive conflict: Conflict that is both productive and beneficial to parties involved. Although disagreement over ideas, decisions, or directions exists, an atmosphere of openness and respect is maintained.

Co-regulation: A supportive process where individuals help each other navigate their emotions and behaviors.

Cultural humility: Lifelong process of self-reflection and self-critique where individuals learn about others' culture and also examine their own beliefs and cultural identities. Understanding and respecting the cultural identities of individuals and communities with an open and respectful stance.

Culture: A concept that encompasses the social behavior, institutions, and norms found in human societies, as well as the knowledge, beliefs, arts, laws, customs, capabilities, attitudes, and habits of the individuals in these groups. "It's the way things are done around here."

Depression: A medical illness characterized by feelings of severe despondency, sadness, and/or inadequacy lasting more than two weeks, often accompanied by lack of energy and disturbance of sleep and appetite and a loss of activities once enjoyed.

Distress (*negative*): The adverse psychological, physical, behavioral, and organizational consequences that may arise secondary to stressful events. *Note: Clinical research confirms that the sense of having little to no control is always a cause of distress.*

Emotional fatigue: Decreased emotional energy as a result of excessive personal or work demands. Emotional exhaustion for caregivers can be a result of the experience of empathy fatigue, moral distress, emotional labor, and chronic caretaker burnout.

Emotional labor: The work of managing one's own emotions and expressions when interacting with others, especially in service, health care, and leadership roles. It often involves suppressing negative emotions like frustration or anger and striving to present a calm, pleasant demeanor.

Emotional regulation: Strategies to pause and assess a potential threat can be helpful in managing an automatic reaction and choosing to intentionally respond to the situation instead.

Emotional resilience: The capacity to adapt and thrive in the face of adversity and stress. It involves handling stress effectively, staying calm during tough times, and managing one's feelings and behaviors in response to challenges.

Empathic strain or distress: Historically categorized as "compassion fatigue," this is a form of secondary stress resulting from the physical, emotional, and psychological impact of helping others and exposure to the pain and distress of others (secondary traumatic stress or STS). Can lead to the diminishment of the empathic response over time.

Empathy: The ability to share and understand the feelings of another.

- *Cognitive empathy*: Ability to understand another person's thoughts, emotions, and perspective intellectually without necessarily feeling them yourself.
- *Emotional (affective) empathy*: The experience of physically or emotionally sharing another person's feelings, creating a deep interpersonal connection. It enables mirroring of emotions, fostering compassion and bonding. Excessive emotional empathy can lead to overwhelm or burnout.
- *Compassionate empathy (empathic concern):* Combines understanding and emotional resonance with the desire to help or to act for another's wellbeing.

Equanimity A state of psychological balance and stability. A willingness to accept things as they are without clinging to or rejecting them.

Equity: Fairness or justice in the way that people are treated. Equity recognizes that people have different needs and challenges and acknowledges disparities in the way people of different races, genders, religions, etc., may be treated.

Eustress (*positive*): Appropriate physiological reaction and psychological response resulting from a motivating, enjoyable, or beneficial challenge.

Fixed mindset: The belief that one's intelligence, abilities, and talents are static and cannot be significantly improved through effort or learning.

Flourishing: Generally, this refers to vigorous and healthy growth. In positive psychology terms, it means feeling happiness, positive emotions, and having purpose and meaning in one's life, all of which contribute to life satisfaction.

Futile care: The continued provision of medical care or treatment to a patient when there is no reasonable hope of a cure or benefit to them.

Growth mindset: The belief that one's abilities and intelligence can be developed and improved through dedication and hard work. Individuals that adopt this frame of mind see challenging circumstances as learning opportunities.

Hardiness: A personality style that helps a person cope, withstand, and actively engage in transformational coping when faced with stressful events.

Happiness: As defined by the positive psychologist, Dr. Lyobumirsky: *"the experience of joy, contentment, or positive wellbeing, combined with a sense that one's life is good, meaningful, and worthwhile."*

Imposter phenomenon: A psychological occurrence in which an individual doubts their skills, talents, or accomplishments and has a persistent internalized fear of being exposed as a fraud. Despite external evidence of competence, those experiencing this phenomenon do not believe they deserve their success or luck.

Inclusion: In the workplace, inclusion is the practice of creating an environment where all people feel valued, accepted, and respected, regardless of their differences. An inclusive workplace fosters a sense of belonging and encourages all individuals to contribute to their fullest potential.

Intentional communication: Communication that involves speaking with awareness and purpose as well as consideration of its potential effects on others. Communicating in a way that is received well and has a positive lasting impact with an individual, team, or across the organization.

Job crafting: An individually driven work design process that refers to self-initiated, proactive strategies to change the characteristics of one's job to better align the job with personal needs, goals, and skills.

Life-work integration: A method of intermixing and synchronizing all aspects of life, from professional to personal, to support health and subjective wellbeing.

Meditation: Utilizing various practices to quiet the mind or to achieve a higher level of consciousness. Meditation is one of the techniques such as deep breathing and body scans to cultivate mindfulness.

Mindfulness: Paying attention on purpose, in the present moment, without judgment. A mental state of nonjudgmental awareness of what is happening in the present moment, including the awareness of one's own thoughts, feelings, and senses.

Moral distress: The emotional state that arises when one identifies the ethically correct, or "right," action to take but feels powerless or constrained from taking that action.

Moral resilience: The capacity of a person to sustain, restore, or deepen their integrity in response to moral complexity, confusion, distress, or setbacks.

Moral resilience is founded on self-knowledge of and commitment to one's values and intentions.

Multidisciplinary approach: The integration of knowledge, methods, and perspectives among professionals from different disciplines to address complex problems, enhance learning, and improve outcomes in various fields such as health care and education.

Mutuality: The sharing of a feeling, action, or relationship between two or more parties. There is good communication and a sense of understanding of the other(s).

Neuroplasticity: The brain's capacity to continue growing and evolving in response to life experiences and intentional practices. Neuroplasticity allows the brain to adapt or change over time by creating new neurons and building new networks.

Non-sleep deep rest: A practice that promotes deep relaxation and recovery without the need for actual sleep, allowing individuals to recharge mentally and physically and providing restorative benefits similar to deep sleep.

Occupational burnout: A workplace syndrome of physical and emotional exhaustion secondary to chronic occupational stress, with a spectrum of severity and symptomology. Distinct from anxiety and depression disorders, but occurrence and symptoms may overlap. Differs from compassion fatigue in that burnout usually stems from excessive responsibilities or time at work and/or inadequate resources to do the work (e.g. staffing, equipment, training).

Occupational distress: The negative psychological and physical effects an employee experiences due to the responsibilities, environment, or other pressures of the workplace.

Occupational wellness: Refers to finding personal fulfillment and meaning through your work while maintaining balance with other life domains. It requires clarity on personal values, purpose, and strengths.

Open-ended inquiry: Questions that invite expansive, more detailed answers rather than simple "yes-or-no" answers, allowing for a deeper understanding of thoughts and feelings. These questions often begin with works like "how," "what," or "why."

Optimism: The tendency to anticipate favorable outcomes.

Pathological altruism: Behaviors that attempt to promote the welfare of others but themselves have damaging long-term consequences for the caregiver.

Peer support: A peer is someone who can offer support based on their relevant life experiences that allows them to relate to others in similar circumstances. Peer support is defined as a way of relating to someone through those shared experiences and commonalities, aimed at offering and receiving support.

Perfectionism: Having excessively high personal standards and overly critical self-evaluations. Common in high-achieving professionals.

Polyvagal Theory: Developed by Dr. Stephen Porges in 1994, this theory is a collection of evolutionary, neuroscientific, and psychological constructs that explain the role of the vagus nerve in emotion regulation, social connection, and the fear response.

Positive leadership: A leadership style that focuses on fostering an uplifting, supportive, and empowering work environment. It emphasizes optimism, encouragement, and the wellbeing of employees. By promoting a positive climate and fostering trust and collaboration among team members, it enhances team morale, productivity, and organizational success.

Positive psychology: The study of what makes life worth living, focusing on concepts such as happiness, resilience, meaning, and strengths.

Presenteeism: The practice of being present at one's work despite health issues, stress, or other personal challenges that prevent someone from performing at their best. There are variety of ways that presenteeism can manifest, each with distinct implications for the individual, team, and the workplace: sickness, mental health, stress or burnout, job insecurity-driven, and overtime or "extra-hours" presenteeism.

Professional satisfaction: An individual's contentment, fulfillment, and positive emotions toward their job. Contributing factors include meaningful work, fair compensation, professional development, and life-work integration.

Protective factors: Characteristics or attributes that reduce the likelihood of negative outcomes and enhance an individual's ability to face adversity. Some examples are supportive relationships, positive self-esteem, and effective coping strategies.

Psychological health: A state of balance that consists of self-acceptance, positive relationships with others, a sense of purpose, autonomy, and competence.

Psychological safety: Employees feel included, safe to learn, safe to contribute, safe to challenge the status quo without fear of being embarrassed, marginalized, or punished

Psychosocial challenges: The difficulties that arise from the interaction between psychosocial factors and social environment, potentially impacting an individual's mental wellbeing and emotional health. Examples include job-related stress, family issues, and social pressures.

Reflective listening: A communication technique that involves actively demonstrating understanding and empathy toward the speaker. It includes repeating or paraphrasing the speaker's thoughts, feelings, or ideas to confirm comprehension and ensure clarity in the conversation.

Resiliency: A multidimensional construct; a personal resource dependent on age, sex, origin, culture, and life circumstances. The psychological capacity to adapt to stressful circumstances and bounce back from adverse events.

Secondary traumatic stress (STS): Physiological and psychological responses following secondary exposure to challenging events or materials even on one occasion. A natural response to knowing about, seeing, or hearing the traumatizing events experienced by others. It can also result from helping or wanting to help those who are suffering or have suffered.

Self-compassion: Offering understanding, acceptance, and kindness toward ourselves when we fail or suffer.

Sleep hygiene: Healthy habits, behaviors, and environmental factors that can be practiced to promote better sleep.

Stigma: Refers to the negative attitudes, unfair beliefs, and stereotypes that a society or group of people have about something, e.g., mental health stigma.

Stoicism: A trait characterized by self-discipline, emotional resilience, and the ability to remain calm and composed in the face of adversity.

Strengths: Stable traits that can be developed and support wellbeing and wise, practical responses to challenges in life and include self-esteem, resilience, positive emotions, distress tolerance, learned optimism, emotional intelligence, compassion, and gratitude.

Stress: Refers to the natural reaction your body has when changes or challenges occur. When stressors are experienced, the body produces physical, mental, and emotional responses. The 3 main kinds of stress: acute, episodic acute, and chronic.

Stress cycle: Refers to the body's response to stressors and stress reactions, which includes 3 stages: alarm (initial reaction to stress – body prepares to respond to the perceived threat), resistance (body attempts to adapt to stressor to maintain homeostasis), and exhaustion (occurs with prolonged exposure, resulting in depletion of resources, fatigue, and decreased performance).

Stress reaction: The unconscious prep to fight, flee, fawn, or freeze.

Stressors: External events, factors, or people that trigger the stress response

Sustainability plan: A personal project that helps you identify and implement the practices that support your wellbeing in different aspects of your life.

Systems-informed positive psychology (SIPP): Integrates systems science principles into positive psychology to optimize human social systems and individual wellbeing. SIPP emphasizes the importance of context, relationships, and dynamic nature of human experiences.

Trauma-informed care: A framework that acknowledges the widespread impact of trauma and integrates this understanding into care practices. It focuses on creating an environment where individuals feel physically and emotionally safe, avoids re-traumatization, and recognizes that individuals may have previous trauma that shapes their current behavior and needs.

Wellbeing: The state comprising of a person's physical, mental, emotional, and social health.

Wellness: The process of taking action to maintain and improve wellbeing.

Work-life integration: A method of intermixing and synchronizing all aspects of your life, from work to personal.

1

The Trail Map

Where We've Been, Where We Are, and Why This Workbook Exists

Orienting Ourselves: An Overview of the Global Veterinary System

The human relationship with animals has existed and has been evolving for thousands of years. The timeline over the last 200 years for animal health and welfare mirrors the roles that animals have played in human lives – industry, food, protection, companionship. Formal training for veterinarians began with the opening of the Royal College in London in 1791. Over the course of the next several hundred years, opportunities for individuals interested in and committed to animal and human welfare have grown globally at different rates, with a wide range of curricula and applications.

The world of veterinary education is both vast and diverse today. It is estimated that there are currently over 200 veterinary schools spread across various countries worldwide. These programs usually last 5 to 7 years, varying by country, and offer students essential knowledge and clinical skills. Veterinary schools differ in size, resources, and standards, but all play a key role in meeting the demand for animal health professionals. There are numerous veterinary technician training programs now worldwide as well. These programs are typically offered at vocational schools, colleges, and universities, leading to various degrees and certificates. Accredited programs ensure that graduates are prepared for certification and licensure as veterinary technicians. There are many more animal health professionals that provide care and

DOI: 10.1201/9781003475361-1

support of clients, the veterinary team, and veterinary patients as client care coordinators, veterinary assistants, kennel staff, hospital administrators, and practice/organization owners and leaders.

My point in bringing this up is that as of today, there is no one way to train to be or to work as a veterinary professional or to support animal welfare. The upside of this is that here is a world of opportunity to participate in the support of a One Health approach to animal, human, and environmental wellness. The challenges are that there is a great deal of variety in approaches and standards around training of animal care professionals and that there are not yet veterinary technicians or nursing training programs in many countries where veterinary medicine is taught and practiced. This matters in that the need for animal health support and services is high and growing. Meeting the needs of the animals, honoring the human–animal bond by supporting animal guardians, and proactively preparing veterinary health professionals and leaders to sustainably meet the innate challenges of veterinary medicine are at the crux of conversations examining veterinary medicine's future.

How do you, and we, meet this pivotal and essential dialogue with more preparedness, awareness, and energy? What professional initiatives and strategies exist or need to be created to attract, intentionally prepare, and proactively upskill veterinary care professionals and leaders to ethically meet the high demands of veterinary medicine today and tomorrow? These are bold and complex inquiries that require both individual fortification and collective commitment to co-create sustainable solutions and strategies. A multidisciplinary approach where wisdom is drawn widely from the veterinary profession, other caregiving professions, and sources of relevant strategy-creation such as positive psychology, social and behavioral science, and business is smart, efficient, and necessary.

If this feels both daunting and exciting, you are in the right place and you are in good company.

What We Have Learned (and Are Still Learning)

In early 2020, the COVID-19 pandemic arrived and shook the world collectively. For veterinary professionals, the disruption went far beyond masks and protocols. It forced many of us to pause, reflect, and ask: *Am I okay? Is this sustainable? What matters most right now?*

The "Great Resignation" – which I prefer to think of as the "Great Reprioritization" – ushered in conversations we needed to have long before. New terminology entered our daily vocabulary: curbside care, remote work, case diversion, and moral fatigue. While fear and uncertainty ran high, one thing remained true: the veterinary profession showed up.

Veterinary professionals – selfless, tenacious, skilled – continued to care for animals, their guardians, and each other. What emerged was a global acknowledgment of the profound emotional, psychological, and physical toll this work takes on the caregivers themselves.

The pandemic amplified work-related stressors that have long existed in veterinary practice, but it also brought new opportunities. We saw:

- A surge in conversations around trauma, grief, burnout, and mental health.
- A rapid expansion of virtual learning and peer support.
- An influx of wellness initiatives from veterinary organizations, finally beginning to address long-neglected needs.
- The utilization of existing as well as the creation of new, veterinary-centric technologies to support the veterinary team in their administrative, communication, and caregiving work.
- The willingness to give ourselves and each other the compassion historically reserved for veterinary patients.

The collective experience deepened our understanding of the realities of caregiving in this profession. It also set the stage for meaningful systemic change.

We learned that wellbeing isn't just about individual resilience, it's about redesigning the environments in which we work.

What Research Shows

Over the course of the last several decades, the amount of research in and around veterinary medicine has flourished. Curricula and support of students and staff is being closely examined, the workflow and support of veterinary workers is growing, and the psychosocial challenges and preparation strategies needed for caregiving professional wellness is being more widely discussed than ever before.

There are many more research efforts and publications that have direct application to the veterinary industry than I will share in this chapter or in the later resource-sharing chapters. I want to pique your curiosity as well as give you both hope and awareness around what has, and is currently, being studied and discussed.

Initially, here are a few studies which confirm what many of us within veterinary practice have intuited:

- Organizational strategies and policy shifts – not just individual coping tools – are essential for sustainable health of professionals and of business.[1–3]
- Examining and naming the core stressors for veterinary professionals; these include financial strain, ethical conflicts, toxic culture, and poor client interactions.[4–6]
- Global veterinary workplace wellness concerns and guidelines are needed to guide the necessary conversations to support new approaches and culture change.[7,8]
- Utilization of data-gathering assessments and tools, such as the ProQOL-Health assessment, can provide insights into which workplace initiatives to improve professional wellness and satisfaction are integral in measuring what is working and what else may be needed.[9]
- Stress-management, emotional intelligence skills, and self-care coping strategies need to be provided to all levels of the veterinary team before and when they enter the workplace to support sustainable worker wellbeing and quality veterinary care.[10,11]
- There are universal systemic challenges in veterinary work, including mismatched workload and resources, lack of flexible scheduling, poor prioritization for rest/sleep, moral distress, and inadequate appreciation of individual workers.[12,13]

These findings all point in the same direction: It's time to shift from awareness to action.

The Realities of Veterinary Stressors

As a result of both the increased veterinary-centric research and the initiatives that have encouraged the sharing of lived experiences, there is now greater clarity around the shared work-related stressors for veterinary professionals worldwide. Here are just some of the most commonly cited – and often compounding – stressors veterinary professionals face:

- **Workload and overtime**, often without sufficient staffing
- **Financial concerns**, including student debt and under-compensation
- **Moral stressors**, like futile care, ethical dilemmas, or toxic culture
- **Client difficulties**, including unrealistic expectations and bullying

- **Social media**, cyberbullying, and public shaming
- **Safety concerns**, both physical and psychological[14]

Systemic factors often make these worse. Institutions designed with a focus on patient outcomes and financial viability frequently overlook the wellbeing of the people delivering that care. All levels of the veterinary team need to be considered when creating policies, protocols, and supportive resources. That also requires that those in the roles of mentoring and leading are hired and trained intentionally to support a values-aligned, purpose-driven culture for veterinary organizations.

Reimagining Veterinary Culture

Envisioning a healthier environment that supports both veterinary professionals and veterinary patients requires examining what elements have worked and are working for us and what we must collectively commit to recognizing and releasing. Some of the more historical problematic cultural issues in veterinary medicine (and other caregiving ecosystems) that deserve to be challenged and changed are:

- **Presenteeism** – showing up even when unwell
- **Pathological altruism** – prioritizing others to your own detriment
- **Toxic toughness** – where vulnerability is taboo
- **Lack of psychological safety** – especially for marginalized individuals
- **Inadequate rest, recognition, and support** – for new parents, caregivers, and everyone in between

None of this will change overnight. But with intention, structure, and support, we can begin. Long-standing, deeply engrained ways of thinking and behavior are tough to shift in any system. Identifying and sharing our individual and collective "why" can provide the fuel we need to bring the sustainable energy and resilience needed to lean into these essential change conversations.

My Professional "Why"

When I began writing my first book, *Creating Wellbeing and Building Resilience in the Veterinary Profession: A Call to Life*, I was working in a novel role as a veterinary peer supporter embedded within a veterinary mental health and wellbeing team. I had the opportunity to work alongside and learn from social workers. Together we listened to our veterinary teams, co-created and delivered training, and held supportive conversations. We didn't know it at the time, but that support system would soon become critical in ways we never could have anticipated. The COVID-19 pandemic effects struck just as this team was finding its footing. The immediate and the subsequent ripples of the pandemic on veterinary team workers' wellness in the organization we supported and those around the world impacted many levels of veterinary practice and functioning. The support of our veterinary caregivers and leaders had to expand, adapt, and meaningfully connect in ways that had not really existed previously. So much learning occurred during this time out of necessity for everyone.

I was more motivated than ever in that cauldron of change to bring awareness to the concerns and opportunities to the veterinary community. We need and deserve the knowledge, compassionate community, and strategies that fortify our capacities to be healthy humans doing hard and beautiful work. Although there were some supportive coping tools/strategies shared in the first book, it was always in my heart to create a practice book to accompany and go deeper into practice building toward professional wellness and overall flourishing. A more structured workbook was envisioned to be the "how" to complement the "what" that was explored in the *Creating Wellbeing* chapters. Conversations with colleagues and leaders worldwide at that time and since have made it clear that there has been a dearth of implementable, relevant tools and strategies to practically meet the needs of the veterinary profession.

Between that time and now, I have invested a sizeable amount of time, energy, and personal finances into my personal and professional growth as a veterinary wellness educator, mentor, coach, and speaker. I have earned certifications in trauma-informed care, mindful meditation teaching, veterinary mentorship, and neuro-informed coaching. While working on my own as well as in aligned community spaces, it has been deeply rewarding and educational to share wellness content that I have learned into conversations with veterinary professionals around the world. In doing so, there is always bi-directional learning, and I feel more prepared and excited to bring this workbook to the larger veterinary audience than ever.

My motivation for writing this workbook is simple but urgent: I want us all to stop merely surviving our careers and start truly *living* within them.

I have lived the chaos of ER practice. I've stayed late to support a colleague in tears. I've felt both the incredible privilege and the weighty responsibility of being a veterinary caregiver. I know how easy it is to lose yourself in service to others. And I know how vital it is that we learn new ways forward – not just for ourselves but for the entire ecosystem of veterinary care. The animals and the humans that love them need us. The environments that we co-exist in need us. The humans that benefit from veterinary involvement in food safety and epidemiology need us.

This workbook is my offering to the veterinary profession I love and the generous people that support the care of animal health and human wellness worldwide. It's a resource for reconnecting with meaning, healing from exhaustion, and navigating the messy, beautiful space where life and work intersect.

How Do I Define "Success"?

- Freedom to be my most authentic self and live a life of stability, opportunity, and holistic wellbeing and health.
- Success is also being "seen" as having something of value to contribute to the conversations around culture change and caregiver wellbeing and being asked to speak/mentor/provide support in a variety of veterinary spaces +/– other caregiving spaces that support animal welfare and possibly environmental/conservation endeavors.
- "Being at choice" – Having choice in what initiatives I invest my time and energy into (autonomy and efficacy).
- Flexibility and freedom created by my own strategic approaches to financial stability.
- Contributing in impactful ways to the healing and sustainable wellness of all beings everywhere.

How Is Having Meaning and Impact Important to Me?

- My core caregiving and teaching/mentoring values are integrity, kindness, and quality. I am a healer, an ally, and an advocate for those that suffer and do not have a voice (all animals, including humans).
- I am at a place in my life where I have the ability to discern and select those opportunities that most closely align with these values.
- My goal is to be of compassionate, impactful service to the good that I know that veterinary medicine delivers to the world now and going into the future.

What This Workbook Offers

This workbook is your trail map as we take on the expedition of a lifetime! Think of each chapter as an aid station – here to fuel your body, mind, and spirit for the road ahead.

Chapters 2–5 are structured similarly as they move through the micro, mezzo, macro systems level of veterinary medicine – individual, team, practice/culture. In each chapter, there are subsections with topics that I felt were aligned with supporting what folks may already know or are not yet aware of that can support:

- How to cultivate sustainable life-work integration
- Practical tools to build resilience and emotional agility
- Ways to foster team engagement and psychological safety
- Strategies to reconnect with meaning, purpose, and fulfillment in your work

Changing behavior is hard – welcome to being human. Even when we know that a shift in mindset or habits could improve our lives, it often feels easier to stick with the familiar, even if it's unsustainable. The truth is that change requires time, intention, and repetition to become "sticky."

But here's the real question: Is the discomfort, exhaustion, or frustration you're experiencing right now enough to motivate you to try something different?

Only you can decide when, what, and how you'll take that first step toward a healthier, more fulfilling life in veterinary medicine. But the good news? You don't have to do it alone.

Use this workbook in whatever manner best supports your learning style and goals: solo, with an accountability buddy, or as a team.

Start at the beginning or jump around. Go to the topics that are most relevant to you. Pause when you need to. Celebrate when you grow.

This Journey Starts with You

A healthier *you* means a healthier *team*. And healthier teams create workplaces where veterinary professionals can thrive – not just produce.

This is how we shift our culture. One person, one practice, one insight at a time.

Let's get geared up. Let's be honest. Let's get hopeful. Let's hit the learning trail – together.

The time is now.

Figure 1.1 Welcome: Setting intentions for the journey ahead.

References

1. Shanafelt, T, Stolz, S, Springer, J, Murph, D, Bohman, B, Trockel, M. (2020). A blueprint for organizational strategies to promote the well-being of health care professionals. *NEJM Catal.* 1:CAT.20.0266. https://doi.org/10.1056/CAT.20.0266.
2. Marshal, K, Ed. (2025). *Diversity, Equity, and Inclusion in Veterinary Medicine*. Hoboken: Wiley.
3. Neill, CL, Hansen, CR, Salois, M. (2022). The economic cost of burnout in veterinary medicine. *Front Vet Sci.* 9:814014. https://doi.org/10.3389/fvets.2022.814104.
4. Holowaychuk, MK, Lamb KE. (2023). Burnout symptoms and workplace satisfaction among veterinary emergency care providers. *J Vet Emerg Crit Care*. 33(2):180–191. https://doi.org/10.1111/vec.13271.

5. Steffey, MA, Griffon, DJ, Risselada, M, Buote, NJ, Scharfe, VF, Zamprogno, H, Winter, AL. (2023). A narrative review of the physiology and health effects of burnout associated with veterinarian-pertinent occupational stressors. *Front Vet Sci.* 10:1184525. https://doi.org/10.3389/fvets.2023.1184525.
6. Volk, JO, Schimmack, U, Strand, EB, Reinhard, A, Hahn, J, Andrews, J, Probyn-Smith, K, Jones, R. (2024). Merck animal health veterinary team study reveals factors associated with well-being, burnout, and mental health among nonveterinarian practice team members. *J Am Vet Med Assoc.* 262:1330–1337.
7. Paton, M, Kalemtzaki, E, Stoewen, D, Hameedunisha, T, Yang, H, Donlin J, et al. (2024). WSAVA professional wellness guidelines. *J Small Anim Pract.* 65:153–175. https://doi.org/10.1111/jsap.13689.
8. RCVS Mind Matters. (2025). *Understanding and Managing Stress in the Veterinary Workplace.* https://vetmindmatters.org/download/new-guide-understanding-and-managing-stress-in-the-veterinary-workplace/?2pdmdl=9572.
9. Alwood, AJ, Ferrentino, DA, Olson, SA, Rodriguez, VI. (2024). The positive impact of daily well-being practices on individual veterinary professionals' professional quality of life self-assessment scores within an emergency and specialty hospital. *Front Vet Sci.* https://doi.org/10.3389/fvets.2024.1381090.
10. Corrigan, VK, Newman, RL, Richmond, P, Strand, EB, Vaisman, JM. (2025). The future of flourishing in veterinary medicine: A systems-informed positive psychology approach in veterinary education. *Front Vet Sci.* 11:1484412. https://doi.org/10.3389/fvets.2024.1484412.
11. Steffey, MA, Griffon, DJ, Risselada, M, Scharf, VF, Buote, NJ, Zamprogno, H, Winter, AL. (2023). Veterinarian burnout demographic: A narrative review. *Front Vet Sci.* 10:1184526.
12. Chapman, AJ, Bennett, PC, Rohlf, VI. (2025). Workplace strategies to reduce burnout in veterinary nurses and technicians: A delphi study. *Animals.* 15:1257. https://doi.org/10.3390/ani15091257.
13. Tomasi, S, Peterson, M, Hale, C. (2023). Could the national academy of medicine's national plan for health workforce well-being work as a framework to improve the well-being of our US clinical veterinary healthcare teams? *J Am Vet Med Assoc.* 262:1–6. https://doi.org/10.2460/javma.23.08.0451.
14. Volk, J, Schimmack, U, Strand, E, Reinhard, A, Vasconcelos, J, Hahn, J, et al. (2022). Executive summary of the Merck Animal Health veterinarian wellbeing study III and veterinary support staff study. *J Am Vet Med Assoc.* 260:1547–1553. https://doi.org/10.2460/javma.22.03.0134.

2

The Path Toward Holistic Wellness

It Starts with a Healthy and "Happy" You

I think that we can all relate to having said at some point: "I just want to be happy" or "I will be happy when . . ."

Happiness is not solely an emotion; it can also be considered a fundamental right. The concept of "being happy" is multifaceted and intricate. Let's briefly examine the essence and misconceptions of "happiness."

Historical philosophical perspectives suggest that individuals often miss out on happiness due to actively seeking it. However, it may be more accurate to assert that the lack of understanding regarding what constitutes happiness – and what does not – can result in feelings of dissatisfaction or sadness. You're happy not because everything is perfect; everything is perfect because *you* are happy. Resilience is a crucial personal attribute for achieving happiness. Empirical research indicates that one's sense of happiness significantly influences both success and life satisfaction. Therefore, happiness should be viewed not merely as a consequence of achievement but as a prerequisite for it.

To summarize, happiness is "not":

- An emotion
- Perfection
- The absence of hardship
- A symptom of doing the right thing
- Guaranteed

DOI: 10.1201/9781003475361-2

A recommended resource on this subject is "The How of Happiness"[1] by Sonja Lyubomirsky, who is renowned for her work in positive psychology. She defines happiness as "the experience of joy, contentment, or positive well-being, combined with a sense that one's life is good, meaningful, and worthwhile." Happiness is energizing and widely pursued as a desirable state. Dr. Lyubomirsky's mentor, Dr. Martin Seligman, posits that the aforementioned dimensions can be intentionally cultivated to enhance subjective wellbeing. According to Seligman et al., those who lead lives enriched with pleasure, engagement, meaning, and a sense of community tend to report the highest levels of happiness.[2]

Given these insights into "being happy," it is important to consider evidence-based strategies that foster happiness and wellbeing. Achieving such states requires deliberate effort to modify habits and mindsets accordingly.

In this chapter, we are going to take a deep dive into individual wellbeing strategies: mindfulness, micro-breaks, self-compassion, equanimity, neurologic wisdom, emotional regulation, stress management, and recovery practices. Research has proven these to be some of the most effective ways to support health, resilience, and one's experience of positive emotions more regularly.

This is the part of the preparation for the veterinary wellness journey that is truly up to you: What do you need? What practices will support you best? Flexible, strong teams are made up of prepared, knowledgeable individuals, so this chapter puts the emphasis on personal responsibility, internal skills, and sustainable self-care habits.

> "Yesterday I was clever, so I wanted to change the world. Today I am wise, so I am changing myself."
>
> Rumi, 13th-century Persian poet

Section 1: Mindfulness Philosophy and Practices

I want to start with one of the most proven and powerful toolboxes for holistic wellbeing: mindfulness.

Mindfulness involves focusing on the present moment with nonjudgmental awareness, curiosity, and acceptance. Through meditation and related practices, it helps you cultivate greater attention and emotional regulation. In the veterinary field, where high emotional intensity and demanding schedules are common, mindfulness serves as a tool to ground, refocus, and restore emotional balance.

There are entire books and courses on mindfulness in all of its vast "flavors" and applications. Here, we will lightly touch on some of the fundamental aspects and skills. It is my hope that you will get curious enough to further explore and deepen your practices outside of the offerings of this workbook.

Definition: Mindfulness is paying attention on purpose to this present-moment with a nonjudgmental awareness.[3]

Why it matters: Veterinary professionals face constant emotional and mental challenges – high-pressure client interactions, loss, and ethical dilemmas. Mindfulness practices help to mitigate compassion fatigue, improve focus, and enhance resilience. Research shows that regular mindfulness can lower cortisol levels, improve emotional regulation, and reduce symptoms of burnout.

Putting it into Action

Practices:

- **Mindful Micro-Breaks:** Intentional 5–10-minute pauses to reset (examples given later in the chapter).
- **S.T.O.P. Method:** Stop, Take 3 deep breaths, Observe using your 5 senses, Proceed.
- **Walking Meditation:** Mindful attention during slow, intentional movement.
- **Mindful Listening:** Active listening with full attention to the speaker.
- **Mindful Transitions:** Pausing and breathing before moving to the next task.

Strategies

- Set reminders for mindful pauses during the workday.
- Practice mindful breathing during high-stress moments (longer exhalation than inhalation taps into the parasympathetic nervous system).
- Incorporate mindful awareness into client conversations by staying present and listening deeply.
- Use apps like Happier, Calm, Headspace, or Insight Timer for guided meditation.
- Establish "micro-moment" habits like 3 deep breaths before entering a client room and/or while placing your hand on the door handle; set your intention of how you want to show up for this family as you enter the room.

- Create visual "nudges" to encourage mindful pauses, Post-its with inspirational words or phrases or photos that evoke calm, connection, or perspective.
- Set intentions for mindful moments to occur at the start of each shift and throughout the course of your work duties.

Phrases to Use

- "Let's take a mindful pause."
- "I'm going to take a moment to breathe and reset."
- "I'm choosing to respond, not react."
- "Right now, it's like this."
- "I'm here in this moment."

Considerations

- Avoid rushing through mindfulness; it requires intentional slowing down.
- Be aware of judgment – mindfulness is about observation, not critique.
- Recognize that mindfulness is a practice, not perfection.
- Avoid multitasking during mindfulness moments; it diminishes effectiveness.
- Create boundaries around mindful moments; they are non-negotiable resets.

Reflections

- How has mindfulness changed your stress response in the last week?
- What micro-mindful moment could you implement this week?
- Who on your team could benefit from practicing mindful micro-breaks?

Workbook Prompts

- Describe a moment at work when you felt overwhelmed. How could mindfulness have changed your response?
- What current practices do you use to stay grounded during high-pressure situations?
- How can you incorporate micro-mindful moments into your daily routine?

Mindfulness-Based Stress Reduction (MBSR)

The intentional structure and practices of Mindfulness-Based Stress Reduction (MBSR) were created by Dr. Jon Kabat-Zinn as a class at the University of Massachusetts Medical Center in 1979. The goal was to bring awareness and evidence-based meditative strategies to ordinary people from all walks of life to help ease anxiety, stress, and mind-made suffering using elements from Buddhist practices. The initial training is an intensive, 8-week-long program with a certified instructor that utilizes a combination of mindfulness meditation, body awareness practices, and simple yoga postures. Over the course of the last 50 years that MBSR has existed, has been taught, and has been integrated into people's daily lives, this approach to mindfulness has grown to include community practicing spaces, ongoing research, and many trained teachers globally.

Mindfulness as taught in MBSR stems from the Vipassana tradition of insight meditation. Of the several forms of Buddhism largely informed by their place of origin, this is one of the main strands that has gained popularity in the West.

MBSR training goals:

- Cultivate self-awareness; respond rather than react to stressors; change habitual reactivity patterns
- Lowering anxiety and reduces the experience of distress when challenging life circumstances arise
- Live with greater ease (less "suffering")[4]

Box Out: The Pain Paradox

Humans generally dislike any flavor of discomfort. The *pain paradox* comes from decades of research and the observations that for most people there is a natural tendency to escape, deny, or withdraw from pain but that in ignoring, numbing, or resisting the current state of discomfort only intensifies and prolongs the distress.[5] "What we resist, the saying goes, persists."[6]

In Laura Woodward's book where she goes more deeply into MBSR as a therapeutic approach supporting mental health, she adds the following to the aims of MBSR:

"Exploration of uncomfortable emotions and physical pain in a safe environment to develop less emotional reactivity to them.

"Acceptance of life's challenges, changes, and losses as natural parts of human existence.

"Non-judgement of our experiences."[7]

Evidence-Based, Effective Micro-Breaks

Mindful Micro-Breaks are brief, intentional pauses (usually under 10 minutes) that allow you to reset, recharge, and re-engage your focus and energy. These moments are not just rest, they are purposeful interruptions designed to restore mental, emotional, and physical wellbeing during demanding workdays.

Why It Works

Veterinary professionals often operate at full tilt, with constant multitasking, high emotional labor, and cognitive fatigue. Micro-Breaks provide:

- A "reset" for your nervous system
- A moment of calm and centering to prevent decision-making fatigue
- A boost in focus, creativity, and productivity
- A way to reconnect with your purpose and team

Micro-Breaks have been shown to reduce stress, enhance performance, and foster sustainable engagement – even when practiced for just 5–10 minutes at a time.

Practices

- **10-Minute "Paws":** Step away from the clinical floor, breathe deeply, and check in with your body and mind.
- **Stretch and Breathe:** Practice a 1–3-minute stretch and breath reset at your workstation (Note: cross-body movement and longer exhalation>inhalation stimulate both the polyvagal and parasympathetic nervous systems, which deliver neurotransmitters that counterbalance cortisol and norepinephrine).
- **"Vitamin N" (Nature Break):** Step outside, or stand near a window, and intentionally observe nature. Functional MRIs demonstrate the huge and immediate impacts on calming our brains and bringing back "big-picture" perspective, offering a reset to the prefrontal cortex.
- **Guided Meditation:** Use an app (e.g., Calm, Insight Timer) for a 5-minute reset.
- **Doodle, Daydream, or Diffuse Think:** Engage in a non-task creative activity that rests your cognitive load.

Strategies

- Set a **self-care alarm** or use break-reminder apps (e.g., Stretchly, BreakTimer).
- Start your shift with a **team intention**: "We support breaks today."
- Use a **"pass the baton" system** for coverage and accountability in taking breaks.
- Have a **"micro-break menu"** visible in the breakroom or treatment area.
- Pair micro-breaks with a **hydration or snack ritual** to nourish your body.
- Create **a mindful pause playlist** of short meditations or calming music.
- Invite a teammate for a **two-minute connection check-in.**
- Keep **a small wellness kit** with a journal, aromatherapy, or visual reminder of your "why."

Phrases to Use

- "Let's take a 5-minute brain boost."
- "I'm going to take a mindful moment to reset."
- "Breaks help me show up better – for my team and my patients."
- "This pause is part of my productivity plan."
- "Time invested now for 5–8 minutes to reset my nervous system will help me stay focused and efficient, so I am more likely to get done with my work tasks and shift on time."

Considerations (Red Flags/Pitfalls)

- **Guilt:** Feeling selfish or lazy for pausing.
- **Culture of martyrdom:** Believing you must power through without rest.
- **Fear of judgment:** Worrying about how others perceive your need for a break.
- **Unrealistic expectations:** Thinking a break must be "long" or perfect to be valid.
- **Lack of modeling by leaders:** If leaders don't take breaks, team members often won't either.

Questions for reflection

- What barriers keep me (or my team) from taking mindful breaks?
- Which micro-break idea am I most curious to try first?
- How can I build a micro-break routine that fits my shift flow?
- Who can be an accountability buddy or champion with me?
- What shift would it take in my mindset or team culture to make these feel "allowed"?

Compassion Practices (Self-Compassion, Grief and Loss, and Caregiver Distress)

Compassion practices are intentional, mindful efforts to bring kindness and care to yourself and others in times of suffering. These include self-compassion, collective grieving, and skills to mitigate empathic distress – crucial for veterinary professionals who face repeated emotional, ethical, and relational demands.

Why It Works

Veterinary professionals are drawn to caregiving because of their empathy, integrity, and compassion. But these same traits can create vulnerability to **compassion fatigue**, **moral stress**, and **empathic burnout** when support systems are absent. Compassion practices:

- Replenish emotional energy
- Reduce shame, self-criticism, and isolation
- Enhance resilience in high-stress situations
- Support ethical caregiving and sustainable professional identity
- Normalize grief and shared humanity in loss experiences

Practices

- **Self-Compassion Break (Neff):** "This is hard. I'm not alone. May I be kind to myself."[8]
- **RAINN Practice (Tara Brach):** Recognize, Allow, Investigate, Non-Identify, Nurture.[9]
- **Visualizing the Compassionate Self:** Imagine yourself acting with warmth, calm, and strength.

- **Grief Journaling:** After patient loss, write freely for 5 minutes – acknowledge feelings without judgment.
- **Micro-Mourning Rituals:** Light a candle, ring a bell, or take a collective pause post-loss.

Strategies

- Include **self-compassion check-ins** during staff meetings or shift huddles.
- Encourage **permission-giving language**: "It's okay to not be okay."
- Integrate **compassionate debriefs** after critical incidents or a challenging euthanasia.
- Place **self-compassion affirmations** in break rooms or treatment areas.
- Offer **team workshops** on emotional regulation and empathy boundaries.
- Role model **emotionally intelligent leadership**, acknowledging team efforts and emotional tolls.
- Support team members in accessing **peer support, Employee Assistance Programs (EAPs), or counseling** without stigma.
- Normalize grief expression with **emotional safe zones** or "quiet corners."

Phrases to Use

- "You don't have to do this alone."
- "I'm here, and I care about what you're feeling."
- "We did our best – and that matters."
- "This hurts, and we can hold space for that."
- "Self-kindness is not indulgent – it's essential."

Considerations (Red Flags/Pitfalls)

- **Pathological altruism:** Overextending beyond what's healthy or sustainable.
- **Self-judgment during grief:** Believing you "should be over it by now."
- **Toxic resilience culture:** Expecting yourself or others to push through without pause.

- **Suppressed emotion:** Avoiding grief or distress can lead to emotional numbing or burnout.
- **Lack of team modeling:** When leaders don't show vulnerability, others won't feel safe to.

Questions for reflection

- Where in my day do I pause to offer myself kindness or compassion?
- How do I process grief or difficult outcomes – and how could that be supported?
- What helps me re-center when I feel overwhelmed by others' suffering?
- Who in my community or workplace can help hold space for my emotional needs?
- What shifts can I make – individually or as a leader – to cultivate a more compassionate culture?

Equanimity and Radical Acceptance

This subsection is dedicated to exploring the powerful mindfulness concepts of equanimity and radical acceptance. These practices are rooted in emotional stability and acceptance of circumstances and are designed to help veterinary professionals and leaders navigate the complexities of caregiving with resilience and emotional balance. By cultivating these practices, professionals can enhance their wellbeing, reduce burnout, and improve their capacity to provide compassionate care.

Definitions

Equanimity

A state of psychological balance and stability, marked by non-reactivity and acceptance of what is without clinging or rejecting. Equanimity allows veterinary professionals to remain composed and compassionate, even amidst stressful situations.

Radical Acceptance

The intentional practice of fully accepting situations that are beyond one's control without resistance. It is not approval or resignation but rather a way to diminish suffering by letting go of the struggle against reality.

Why These Practices Matter

For veterinary professionals, daily exposure to high-pressure environments, emotional cases, and unpredictable outcomes can be profoundly taxing.

Integrating equanimity and radical acceptance into mindfulness practices empowers individuals to:

- Maintain emotional balance and clarity in the face of distress.
- Navigate compassion fatigue and prevent burnout by releasing attachment to outcomes.
- Strengthen decision-making and interpersonal communication with colleagues and clients.
- Reduce emotional reactivity and remain grounded during difficult interactions.

Equanimity Practices

1. **Impermanence Practice**
   - Focus on the transient nature of emotions and experiences.
   - Observe sensations without attachment; recognize that everything passes with time.
2. **Sphere of Influence Practice**
   - Distinguish between what you can and cannot control.
   - Direct your energy toward manageable elements and release what is beyond your influence.
3. **Non-Clinging Awareness Practice**
   - Stay present with both pleasant and unpleasant experiences without reacting.
   - Label emotions as "pleasant" or "unpleasant" and practice observing without judgment.
4. **Reframing Practice**
   - Explore alternative perspectives when caught in emotional reactions.
   - Consider different interpretations and choose the most balanced viewpoint.

Strategies

1. **Cultivate Self-Awareness:** Regularly check in with your thoughts and feelings without judgment.
2. **Mindfulness Meditation:** Engage in meditation to build a calm, centered mindset.
3. **Reframe Challenging Situations:** View obstacles as opportunities for growth.
4. **Develop Compassion:** Practice empathy toward yourself and others to maintain perspective.
5. **Prioritize Self-Care:** Sustain physical and mental wellbeing to support emotional resilience.

Phrases to Use

- "This too shall pass."
- "Right now, it's like this. And it will pass."
- "I can't control this, but I can control my response."
- "I choose to respond with calm and clarity."
- "I accept the present moment as it is."
- "May I experience safety, ease, and health. May all beings experience safety, ease, and health."

Considerations (Potential Barriers and Red Flags)

1. **Confusing equanimity with indifference:** Equanimity is about emotional stability, not detachment.
2. **Striving for perfection:** Avoid the expectation of flawless calmness; the goal is to grow incrementally.
3. **Over-identifying with emotions:** Acknowledge feelings without becoming overwhelmed by them.
4. **Misinterpreting acceptance as passivity:** Radical acceptance is about acknowledging reality, not surrendering to it.
5. **Resisting discomfort:** Acceptance includes embracing discomfort as a part of growth.

Reflections: Next Steps

1. Identify a recurring stressor and apply equanimity and radical acceptance practices.
2. Journal your observations on how these practices change your emotional response.
3. Practice with a colleague or peer, supporting each other in mindful reflection.
4. Assess your self-care practices and adjust to better support emotional stability.
5. Commit to small daily mindfulness exercises that strengthen these principles.

Section 2: Neurologic Wisdom and Our Human "Operating System" *("Did This Thing Come with a Manual?")*

Neurologic wisdom is the understanding and intentional engagement with how our brain and nervous system respond to stress and learning. It includes principles of neuroplasticity, cognitive reframing, emotional regulation, and the growth mindset. Veterinary professionals can use these evidence-based practices to build resilience, emotional awareness, and adaptability in high-pressure environments. The knowledge and skills gained through practice are highly applicable to all aspects of our lives, inside and outside of the work environment.

Definitions

Neuroplasticity

The brain's capacity to reorganize and form new neural connections throughout life. Through intentional practice, repetition, and time, we can literally change our minds – optimizing emotional regulation, enhancing learning, and supporting stress management.[10]

Cognitive Reframing

A technique used to shift perspective by altering how we interpret situations. This allows for more constructive responses, reducing anxiety and fostering resilience.

Growth Mindset

A belief that abilities and intelligence can be developed through dedication and hard work. This mindset encourages resilience, learning from mistakes, and adapting to new challenges.

Emotional Regulation

The ability to understand and influence one's own emotional state, utilizing supportive skills and practices to maintain emotional wellbeing and adapt to varying situations. Techniques include understanding triggers, working within one's window of tolerance, and using methods like the Emotion Wheel for emotional awareness and processing.

Why These Practices Matter

The veterinary field is mentally and emotionally demanding. Understanding how your brain adapts and processes stress allows you to work with your natural responses rather than against them. Building emotional regulation through neuroplasticity and cognitive reframing improves decision-making, reduces burnout, and strengthens interpersonal relationships.

For veterinary professionals, the demands of caregiving, client interactions, and unpredictable outcomes can lead to emotional strain and burnout. Harnessing neurologic wisdom through these three key approaches can:

- Enhance emotional resilience by rewiring habitual responses to stress.
- Foster adaptive learning, turning setbacks into growth opportunities (I often jokingly call the challenging moments of life "unscheduled growth opportunities" to reframe in the moment).
- Strengthen decision-making and problem-solving under pressure.
- Support mental health and wellbeing by reducing anxiety and improving emotional regulation.

Neuroplasticity Practices

1. **Intentional Focus Practice**
   - **Activity:** Choose a daily mindful focus (breathing, sensation, or affirmation) and track your ability to return to it throughout the day. Log distractions and reflections in a journal. Practicing this regularly strengthens neural pathways for attention and emotional regulation.
   - **Reflection Questions**
     - "What patterns do I notice when I get distracted?"
     - "What sensations arise when I focus intentionally?"
   - **Veterinary Scenario:** During a high-pressure emergency case, you focus on your breath to maintain calm while making critical decisions, for example, using the box-breathing technique.
2. **Mindful Attention to Stress Response:**
   - **Activity:** Recognize and identify your triggers that activate stress responses (fight, flight, fawn, or freeze). Create a "Response Map." Take the physiological sigh method during moments of stress to activate the parasympathetic nervous system. Document your experience.
   - **Reflection Questions**
     - "How did grounding affect my response to stress?"
     - "What surprised me about my triggers?"
   - **Veterinary Scenario:** When facing a difficult client, you use grounding techniques to stay calm and maintain professionalism.

3. **Incremental Learning and Exposure**
   - **Activity:** Gradually expose yourself to stressful scenarios with mindful presence to build tolerance and adaptive response. Create a "Challenge Ladder." List progressively challenging situations and practice mindful exposure weekly.
   - **Reflection Questions**
     - "What small successes did I achieve?"
     - "What adjustments can I make for next time?"
   - **Veterinary Scenario:** Gradually increase exposure to handling complex surgeries with mindful awareness.

Box Out: Ladder of Competence

As we learn new ways of doing and being, there is the opportunity to also climb the "ladder of competence." Initially, we may start at State 1, Unconscious Incompetence, when we do not know what we do not know. Perhaps with time, we may move to Stage 2, Conscious Incompetence, where we become aware of what we do not yet know. And if it is important to us to learn something new, we practice. With this practice, intention, and adapting when mistakes are made, we move forward to Stage 3, Conscious Competence. It still feels awkward, but we are on our way!

Over time, we continue to do this new thing until it becomes less effortful. This is Stage 4, Unconscious Competence. This "new" way of doing or being becomes a habit; it becomes second nature to you.[11]

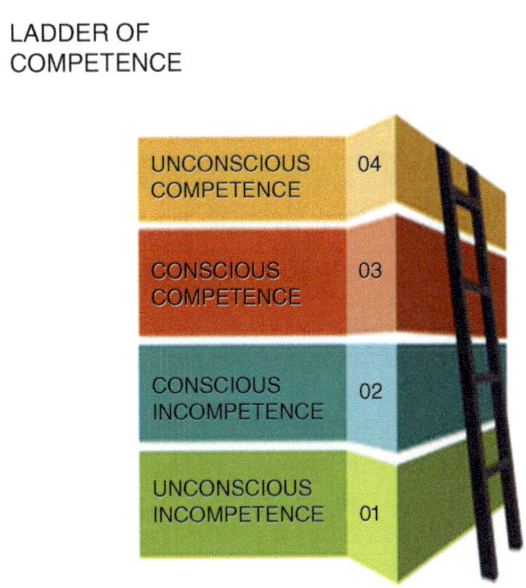

Figure 2.1 The "Ladder of Competence" reflects the stages of growth and learning.

Cognitive Reframing Practices

1. **Perspective Shifting Practice**
   - When faced with a challenge, intentionally reframe the situation by identifying potential learning or growth opportunities.
   - Ask reflective questions like: "What can I learn from this?" or "How can I grow from this experience?"
2. **Rule of Six Technique –** the Native American practice to expand possible theories we may hold about an event or something that is challenging us
   - When automatic negative thoughts arise, identify six alternative ways of interpreting the event.[12]
3. **Narrative Re-Authoring**
   - View your life as a narrative. Reframe disempowering stories as opportunities for growth and resilience.

Growth Mindset Practices[13]

1. **Circle of Control and Influence**
   - Focus your energy on what you can control and release attachment to what you cannot (see Figure 2.2).
2. **Failing Forward**
   - Recognize mistakes as part of the learning process.
3. **Curiosity-Driven Reflection**
   - Approach new situations with curiosity rather than judgment.

Emotional Regulation Practices

Name It to Tame It Using Plutchik's Wheel of Emotions[14]

(*Note: Please also refer to the section on "Emotional Regulation"*)

Use the Emotion Wheel to identify and label what you are feeling. Naming emotions reduces their intensity and promotes clarity.

Window of Tolerance Framework[15]

Understand your emotional range for optimal functioning. Identify when you are in hyperarousal (anxiety, overwhelm) or hypoarousal (numbness, shutdown) and apply grounding techniques to restore balance.

Grounding and Somatic Sensing

Utilize the 5–4–3–2–1 method described later in this section, breathing techniques (4–7–8, physiologic sigh, box breathing), and somatic

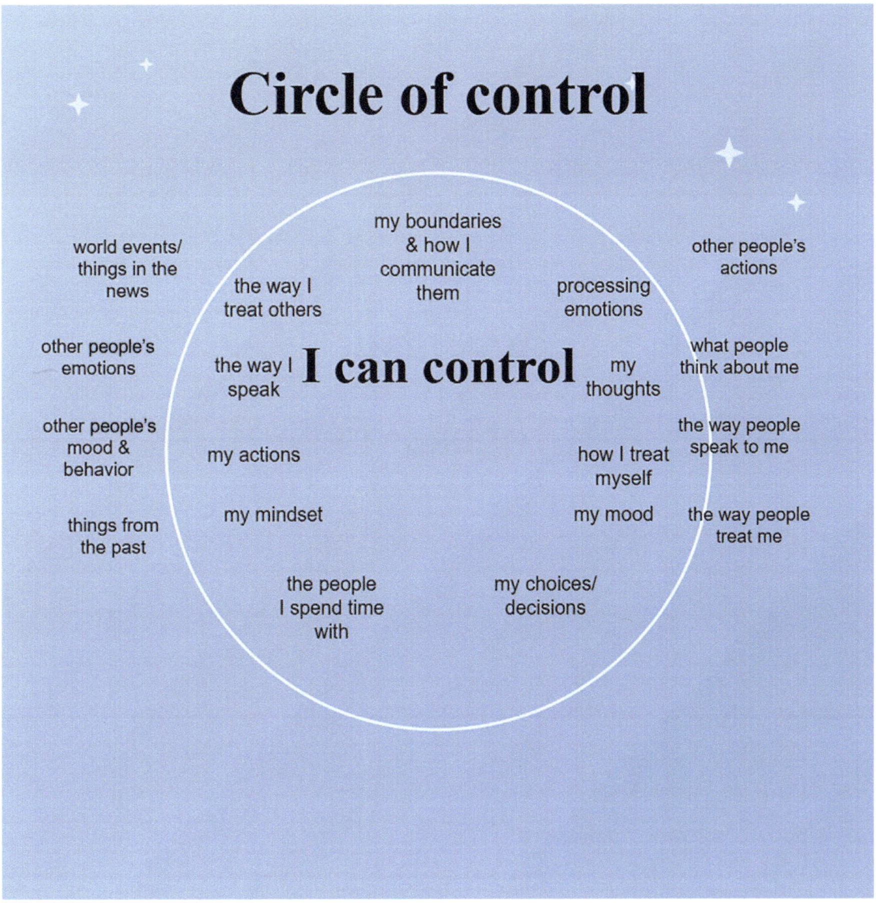

Figure 2.2 Circle of control: For managing energy effectively.

awareness to reconnect with the present moment and regulate emotional states with more awareness and preparation.

Centering Breath

This is more than just taking a deep breath. It's an intentional, on-purpose pause where you reconnect with your body, your awareness, and your center – even for just a few seconds. Taking a series of 3–5 centering breaths can shift you, especially in moments where you might otherwise automatically react.

Here's how:

- Take one slow, conscious breath.
- As you breathe, quietly say the word "center" to yourself.
- Feel your body supported by the ground or chair.
- Let your attention gather into a sense of yourself – grounded, aware, here, in the now (this is somatic sensing).

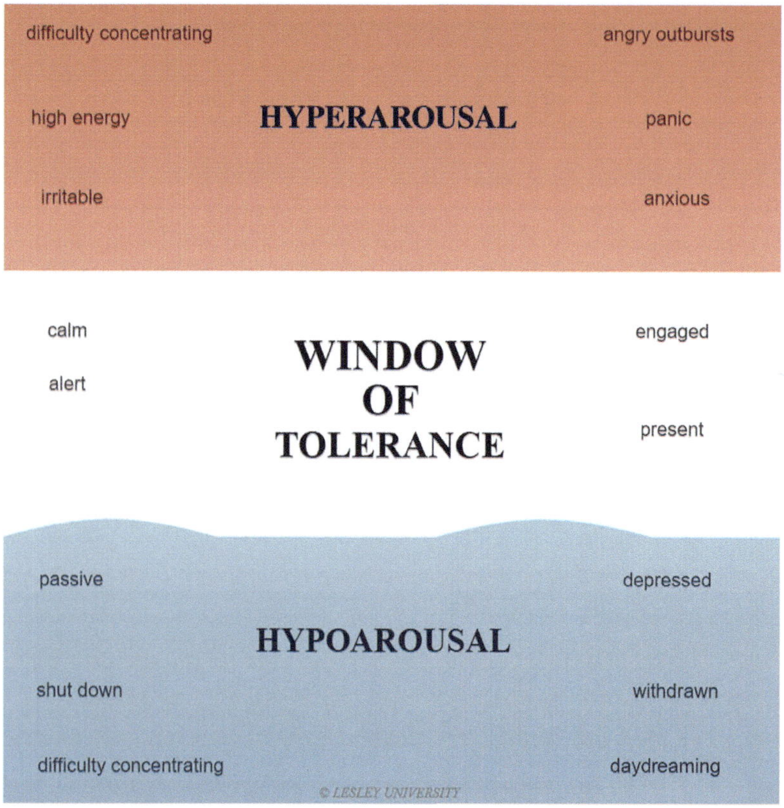

Figure 2.3 Window of tolerance framework.

- Try the 4–7–8 breathing approach to activate your parasympathetic nervous system: Breathe in deeply for 4 counts, hold your breath gently for 7 counts, and slowly exhale through your mouth using your abdominal muscles for 8 counts.

Phrases to Use

- "This is an opportunity to grow."
- "I'm learning through this experience."
- "I can navigate this moment with calm and clarity."
- "I'm building resilience every day."
- "I am aware of my triggers and can manage them."

Considerations

- Avoid striving for perfection; focus on incremental growth.
- Recognize when emotional overwhelm is a signal to pause and ground.

- Avoid avoidance behaviors; incremental exposure builds resilience.
- Be mindful of over-identifying with negative thoughts.
- Understand that emotional regulation is a practice, not an instant fix.

Reflections

- What emotional patterns are you beginning to notice in high-stress situations?
- How has cognitive reframing helped you manage a recent challenge?
- What grounding techniques have been most effective for you?

Workbook Prompts

- Reflect on a recent emotionally challenging experience. How did your brain and body respond?
- What small change could you make to support a growth mindset in high-pressure situations?
- Describe a moment when you successfully reframed a difficult situation. What was the outcome?

Emotional Regulation + Emotional Resilience = Emo-Resilience

Plutchik's Emotion Wheel[14] is a visual tool to help identify and articulate emotions more accurately, which supports emotional awareness and regulation.

Why This Matters

Understanding the spectrum of emotions supports the ability to name and process feelings more effectively, reducing emotional overwhelm. When we can name the emotional experience we are having, we can better understand our current experience with more objectivity and compassion. Having a clearer understanding of what you feel, how you behave, and where your thoughts go with these different emotions gives you the capacity to increase any supportive practices or skills you may need to acknowledge the emotion as data, to learn from it, and then to release it with more skill and efficiency.

If you want to learn more about "how emotions are made" and the research that has been done on emotions and neuroscience, psychologist and neuroscientist Dr. Lisa Feldman Barrett's work is incredibly interesting (https://lisafeldmanbarrett.com).

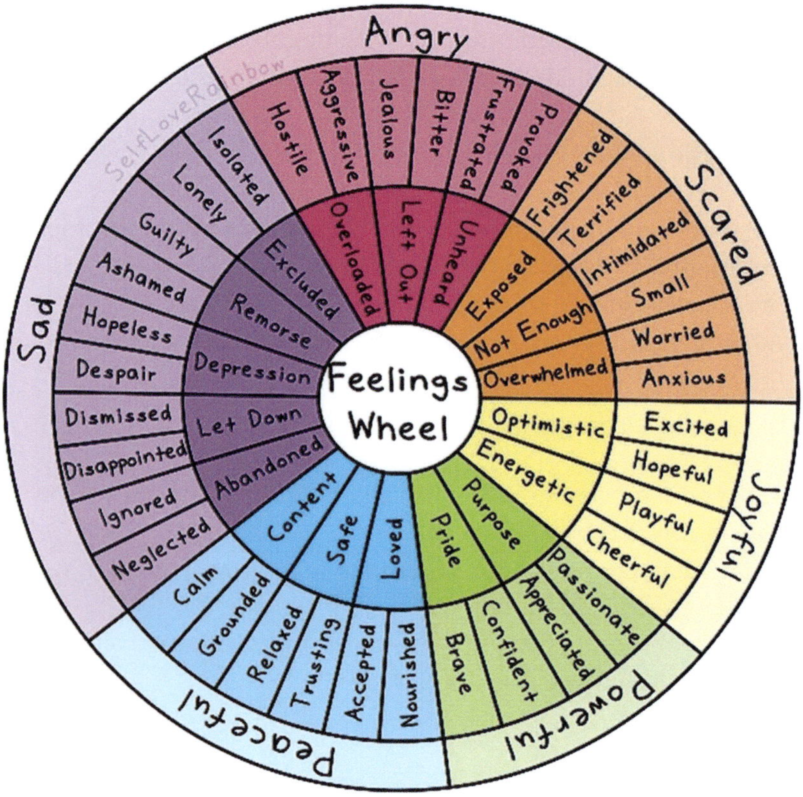

Figure 2.4 Plutchik's emotion wheel.

Practices

- **Emotion Identification:** Using the Emotion Wheel to pinpoint specific emotions.
- **Mapping Emotional Triggers:** Recognizing what events or thoughts lead to certain emotional states.
- **Reflection and Reframing:** Analyzing emotions to find healthier responses.

Questions for Reflection

- How does identifying your emotion change your experience?
- What new awareness did you gain from mapping emotional triggers?
- How can this practice enhance your emotional regulation during stressful moments?

Strategies

1. Build resilience through practice and repetition (supporting the neuroplasticity process)
2. Leverage the "WTF" Framework (Window of Tolerance Framework)[15]
3. Set intentional focus
4. Practice reframing
5. Support systems mindset

Phrases to Use

- "This is a learning opportunity."
- "What can I learn from this challenge?"
- "I am building resilience every day."
- "I choose to respond with calm and clarity."
- "I control my response, not the situation."

Considerations (Potential Barriers and Red Flags)

1. Over-identifying with emotions
2. Expecting instant results
3. Perfectionism and fixed mindsets
4. Avoiding discomfort
5. External validation dependency

Reflections: Next Steps

1. Set a daily intention
2. Observe and reflect
3. Apply the Circle of Control
4. Journal learnings
5. Embrace curiosity

Guided Journal Prompts

1. What emotions frequently arise in your work, and how do you typically respond?
2. Describe a time when shifting your perspective made a significant difference in your emotional response.

3. What grounding technique works best for you in stressful veterinary scenarios?
4. How do you know when you are nearing your window of tolerance? What practices help you return to balance?
5. Reflect on a professional challenge. What did you learn, and how did it shape your approach moving forward?

Self-Assessment Checklist

- I can identify when I am in a state of hyperarousal or hypoarousal.
- I know at least three grounding techniques that work for me.
- I practice cognitive reframing regularly when faced with challenging situations.
- I use the Emotion Wheel to articulate and manage my feelings.
- I actively seek to learn from mistakes and embrace a growth mindset.

Action Plan

1. Identify your top three emotional triggers.
2. Practice one new grounding technique daily for a week.
3. Reflect on a recent challenge and write down three lessons learned.
4. Create a personal mantra for stressful situations (e.g., "Breathe and proceed").
5. Share your progress with a peer or mentor for accountability.

Compassion Resilience and Emotional Agility

Compassion resilience is the capacity to maintain emotional energy and effectiveness in caregiving roles over time, without becoming overwhelmed by the emotional weight of others' suffering. Emotional agility, as defined by Dr. Susan David, is the ability to approach one's inner experiences, thoughts, emotions, and self-talk – with curiosity, acceptance, and courage. Together, these concepts build a foundation for sustainable empathy and emotional strength.[16]

Why It Matters: In veterinary care, emotional strain is inevitable. Daily interactions with clients and patients often involve intense emotional experiences. Compassion resilience allows professionals to extend empathy without depleting their emotional reserves, while emotional agility fosters adaptability in the face of emotional challenges. Building these skills prevents compassion fatigue, enhances team dynamics, and strengthens relationships with clients.

Practices:

- **Compassionate Boundaries:** Setting limits without guilt.
- **Self-Compassion Practices** (guided meditations, kind self-talk).
- **Emotional Flexibility Exercises** (recognizing emotions without judgment).
- **Practicing Empathic Listening** while maintaining personal grounding.
- **Journaling Emotional Insights:** Reflecting on tough days to build understanding.

Strategies

- Establish decompression routines after emotionally intense cases.
- Practice detachment through self-reflection – acknowledge emotion without absorbing it.
- Set limits on emotional labor; know when to step back and reset.
- Engage in regular peer support or debriefing sessions.
- Use "Name It to Tame It"[1] for emotional clarity.
- Implement self-compassion breaks throughout the workday.
- Learn to ask for support when emotional fatigue is setting in.
- Create a personal "Emotional Resilience Plan."

Phrases to Use

- "I can care deeply and protect my own wellbeing."
- "I honor my emotions without being overwhelmed by them."
- "It's okay to step back and recharge."
- "I'm choosing to respond with strength and grace."
- "I can be present for others while caring for myself."

Considerations

- Avoid absorbing others' emotional burdens. Practice healthy detachment and self-awareness.
- Be mindful of compassion fatigue signals; they are a cue to rest and reflect.

- Avoid self-criticism when setting boundaries; it is a form of self-care.
- Recognize the difference between empathy and emotional exhaustion.
- Avoid isolation – engage in community or peer support when feeling depleted.

Reflections

- How has setting emotional boundaries helped you recover your energy?
- What new practice can you implement this week to build emotional agility?
- How would your professional life change if you prioritized compassion resilience?

Workbook Prompts

- Reflect on a recent situation where you felt emotionally drained. How did it affect your work and wellbeing?
- How do you currently manage emotionally intense moments at work? What could you do differently?
- Identify one area where you want to build more emotional agility. How would it impact your practice?

Stress Better: Understanding and Navigating Stress More Skillfully

"Stress Better" is a framework I designed intended to equip veterinary professionals with practical tools and insights to recognize, manage, and thrive in the face of occupational stress. Understanding the physiological and psychological components of stress allows for strategic intervention, emotional regulation, and resilience building.

Why It Matters: Working in veterinary medicine, people (caregivers and clients) can be exposed to high-stress environments, ethical challenges, and emotional strain. Developing skills to "Stress Better" not only enhances individual wellbeing but also strengthens team dynamics and client interactions. Evidence-based strategies help mitigate compassion fatigue, reduce burnout risk, and improve mental clarity during high-pressure moments. Recognizing the difference between distress and eustress allows professionals to leverage stress as a motivator while managing negative impacts effectively.

Concepts and Language Leveling

- **Stress:** The unconscious preparation to fight, flee, fawn, or freeze when facing perceived challenges.
- **Stressor:** External events, factors, or people that trigger the stress response.
- **Distress (Negative):** Adverse psychological, physical, behavioral, and organizational consequences from stressful events.
- **Eustress (Positive):** A positive physiological and psychological response from motivating or enjoyable challenges.
- **Stress Response Cycle:** The biological and psychological processes triggered by stressors, involving a cascade of neural and physiological reactions.
- **Stress First Aid (Psychological First Aid):** A peer-support model to improve recovery from stress reactions in oneself and coworkers.
- **Polyvagal Wisdom:** Understanding of the vagus nerve's role in emotional regulation, stress response, and resilience.

Stress Response Cycle

The Nagoski stress cycle, introduced by Emily and Amelia Nagoski, emphasizes the importance of completing the stress response cycle to prevent burnout and emotional exhaustion.

The stress response cycle is a natural process that occurs in our bodies when we encounter stress. It consists of several stages, including the initial stressor, the physiological response, and the eventual return to a state of calm. However, many people get "stuck" in the cycle, leading to chronic stress and burnout. The Nagoskis highlight that completing this cycle is crucial for emotional and physical wellbeing. In their book, *Burnout: The Secret to Unlocking The Stress Cycle,*[17] the authors offer several common reasons folks get stuck in the stress cycle: chronic stressors causing a chronic stress-activating situation, social appropriateness, choosing to feel "safe" even when it does not deal with the stress itself, and the freeze response as one of the sympathetic nervous system's reactions to stress.

Stress Cycle Completion Matters

When we experience stressors, our bodies react with a series of psychological changes designed to keep us safe, to respond to threats (real or perceived). If we do not complete the cycle by processing emotions and physical sensations associated with the stress reaction, we can suffer a variety of health concerns including anxiety, depression, and chronic pain.[18]

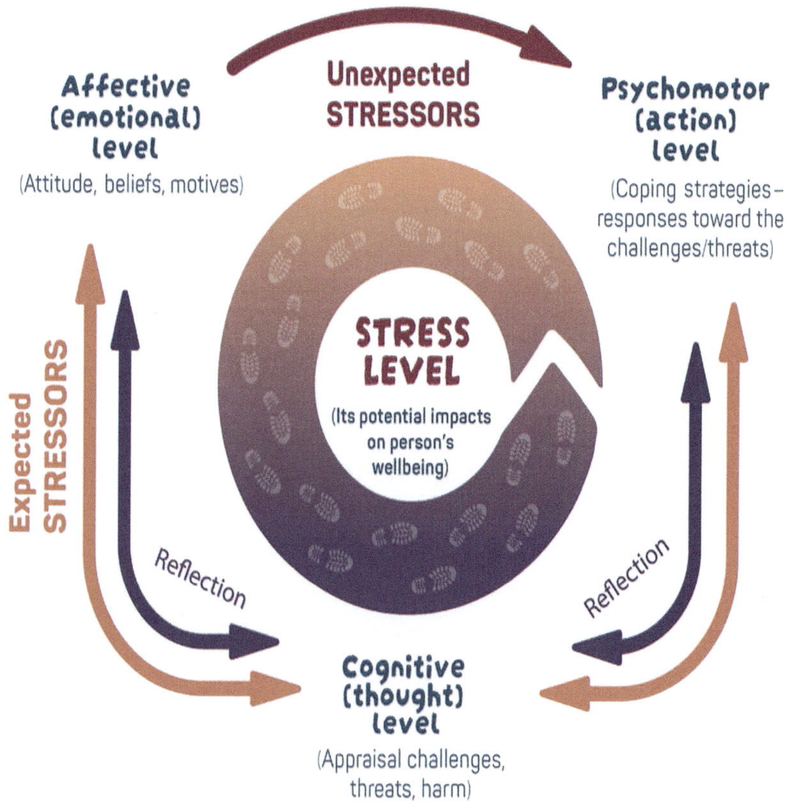

Figure 2.5 The stress response cycle.

Key Takeaway:

You have to deal with the stress itself to "complete the cycle," or there can be long-term negative impacts on your health and quality of life. Per the Nagoski sisters, "Wellness is the freedom to move fluidly through the cycles of being human. Wellness is thus not a state of being; it is a state of action."[17]

Stress First Aid (SFA): Psychological First Aid

Stress First Aid (SFA), also known more globally as Psychological First Aid, is a peer-support model designed to help individuals and teams recognize, address, and recover from stress reactions. It is based on the Stress Continuum Model, which classifies stress responses from "Ready" (Green) to "Reacting" (Yellow), "Injured" (orange), and "Ill" (Red).

This model is shared in this chapter and in Chapter 3 as it applies to stress management awareness and strategies for individuals as well as for

Figure 2.6 Activities to complete the stress cycle.

co-workers/teams. The goal is to normalize that stressors occur in our lives and that stress reactions are common for people. There are a variety of manifestations and sequelae depending upon the individual and how prepared they and their community are in moving through the potential stress reaction impacts. How to recognize, articulate, and support yourself (and perhaps others) through the reaction to return to a state of homeostasis, capability, and functioning and health is the goal of this model.

Why This Matters

Veterinary professionals often face high-pressure environments that trigger acute and chronic stress. SFA provides a structured method to support oneself and others, promoting recovery before stress reactions escalate into larger problems that may impact health, relationships, or career stability. Stress First Aid (Psychological First Aid) provides a structured approach to self-care during and after high-stress situations, reducing burnout and fostering a healthier team culture.[19]

Practices

- **Recognize the Signs**
  - Identify symptoms early (e.g., emotional reactivity, withdrawal, agitation).
- **Provide Support**
  - Validate experiences and listen actively.
  - Offer resources and practical support when needed.
- **Encourage Recovery**
  - Promote self-care practices and peer support.
  - Create an environment that fosters psychological safety.

Strategies

- **The Stress Continuum Model**: Assessing where someone is on the spectrum of stress.
- **Early Intervention:** Addressing yellow zone reactions before they escalate to orange zone and red zone reactions, which are more distressing, impactful, and require more time and resources to resolve.
- **Peer Support Training:** Teaching team members to recognize and respond supportively.
- **Creating Safe Spaces:** Establishing environments where emotions can be shared openly.
- **Routine Check-Ins:** Regularly assessing emotional wellbeing among staff.

Phrases to Use

- "I'm here to support you, how can I help?"
- "I see you're going through a lot; let's talk it through."
- "I recognize that this situation is really challenging."

Considerations

1. Avoid minimizing someone's stress reaction.
2. Be aware of compassion fatigue in peer-support roles.
3. Ensure that interventions are voluntary and consensual.

Reflections

- Consider a recent stressful event – what phase of the stress continuum were you in (green, yellow, orange, red)?
- Who supported you during that time, and what was helpful?
- How might you extend SFA principles to colleagues or team members?

Polyvagal Theory

The vagus nerve is also known as Cranial Nerve X, or the "wandering nerve." This amazing highway of information connects the muscles of the face and inner ear, throat, upper chest, heart area, upper stomach, and lower stomach. Over the decades of research, it has been come to understand that it is helpful to consider the evolution of humans and thus the need for intuitive, subconscious "knowing." Thus, there are "dorsal" and "ventral" aspects of this nervous pathway. Approximately 90% of the "traffic" flows upwards from the gastrointestinal tract and heart area to the brain (particularly the left hemisphere), giving the brain a sense of what the body senses. The vagus nerve plays a role in voice tenor/tone, how we hear, and how the small muscles of our face respond to our environment and other beings, especially humans.

Definition

The Polyvagal Theory, developed by Dr. Stephen Porges, explains how the autonomic nervous system (ANS) regulates our reactions to safety and threat. It focuses on the role of the vagus nerve, which connects the brain to major organs and influences emotional regulation, social connection, and stress response. Understanding that this is not awareness that comes from perception, Porges coined the term *neuroception*.[20] Dr. Daniel Siegel has also written on studies of the vagus nerve and its capacity to be our internal radar. When we listen to our "gut instinct," this is *enteroception* or *interoceptive awareness*. Some of these messages are through the vagus nerve, some are neurochemical transmitters directly from the gut, and research is revealing more about the gastrointestinal tract as a "second brain" every day![21]

Why This Matters

Understanding the Polyvagal System empowers veterinary professionals to recognize physiological responses during stress, enabling more effective self-regulation and emotional resilience. It provides a framework for identifying when one is in a state of fight/flight, freeze, or social engagement. By engaging the vagus nerve through intentional practices, individuals can shift from survival states to a more connected and calm presence. Polyvagal understanding enhances the capacity to self-regulate, promoting mental resilience and wellbeing. This is what some individuals have named "bio-hacking."

Practices

- **Vagal Nerve Stimulation Techniques**
  - Deep 4–7–8, diaphragmatic breathing (inhale for 4 counts – hold for 7 counts – exhale slowly through your mouth for 8 counts)
  - Humming, chanting, or singing (vibrates the vagal nerves in your throat and face)
  - Gentle neck stretches and mindful movement (particularly cross-body movements)
- **Grounding Techniques**
  - **5–4–3–2–1 exercise:** Name 5 things you can see, 4 you can touch, 3 you hear, 2 you smell, and 1 you taste.
  - ***"Feet on the Floor":*** Focus attention on the sensation of your feet firmly planted.
- **Mindful Awareness Practices**
  - Breath work (Box Breathing; noting – "in," "out," and where it is sensed most strongly).
  - Body scanning (starting at your head or your toes, notice and evaluate how the different parts of your body are feeling slowly and with compassionate, nonjudgmental awareness).
  - Progressive Muscle Relaxation (PMR) to release tension and ease anxiety. Clenching and releasing isolated muscle groups one at a time. PMR helps to provide stress relief, decreases symptoms of anxiety, and supports cognitive capacity.
  - Visualization techniques to create a sense of safety.

Box Out: The Stress Breath Exercise to Activate the Polyvagal System

1. Inhale nice and deep, feeling the inbreath move through the back of your throat.
2. Hold your breath and bring your chin down to your chest.
3. Count back from five.
4. Exhale AUDIBLY (perhaps making a *haaaaaa* sound as if you are fogging a mirror).
5. Lift your head back up.
6. That is one cycle. Do twelve in a row, if you can, during the day and again at nighttime.

Why the Stress Breath Technique Works

When the breath is audible on the inhale and on the exhale, vibrations form the sound signal the vagus nerve (the connection between your body and your mind) supports a shift from the autonomic nervous system from the sympathetic (stress response) to parasympathetic (restorative response). This is also called the "physiologic sigh."

Strategies

- **Sequential Relaxation:** Moving from head to toes, intentionally relaxing each muscle group.
- **Mindful Transitioning:** Practicing awareness during transitions (e.g., leaving a stressful event and entering a safe space).
- Create a **"Stress Better"** action plan for high-pressure days.
- **Map out common triggers** and identify proactive coping strategies.
- Use **visualization techniques** for calmness before challenging tasks.
- **Set boundaries** around critical recovery time after stressful events.
- **Implement recovery rituals** – mindful walks, journaling, or reflective pauses.
- **Develop a checklist** for managing acute stress: pause, assess, respond.
- Establish **self-compassion practices** for times of high emotional intensity.
- **Leverage peer support** and collaborative debriefs after difficult cases.

Phrases to Use

- "Let's just be present for a few moments – no need to fix anything right now."
- "I am present and capable in this moment."
- "I acknowledge my stress and can manage it with intention."
- "I am choosing a calm response."
- "I am grounded and resilient."
- "I can step back, breathe, and respond with clarity."

Considerations

1. Avoid ignoring the signs of chronic stress – early intervention is crucial.
2. Recognize when temporary stress becomes prolonged; seek support if needed.

3. Avoid perfectionism; effective stress management is about progress, not perfection.
4. Steer clear of isolation during high-stress periods; connection matters.
5. Be mindful of avoidance behaviors that can mask deeper issues.
6. Avoid over-exerting relaxation techniques when in hyperarousal. Start small.
7. Recognize when professional support is needed if self-regulation does not restore balance.

Questions for reflection:

- Reflect on moments of high stress – what signals did your body give you?
- What practices helped you return to a sense of calm?
- How might you incorporate polyvagal practices into your daily routine?
- What proactive steps can you take to manage your current stressors?
- Who can you reach out to for support when stress becomes overwhelming?
- How can you integrate micro-moments of recovery into your daily routine?

Neuroplasticity and Vertical Brain Integration

With recent technological advances and research endeavors of dedicated neurobiologists, psychologists, behavioral scientists, and other fields of science over many decades, the understanding of the human brain's evolution and function is much clearer and more accurate today. Here are a just a few fun tidbits to consider:

- Your brain is an interconnected network of systems
- The average human brain has a network of 128 billion neurons
- The triune brain idea and that there is a "battle" between emotion, instinct, and rationality is a myth.
- There is no such thing as a limbic system dedicated to emotion.[22]

We now have a much richer view of how our brain's network of neurons takes in, processes, shares, and translates internal and external sensory information. There is indeed a "limbic circuitry," but it is time to remove the outdated language around reptile vs mammal brain and around "limbic hijack."

Humans are a more complicated and interesting animal than that (and so are other animals, by the way, as their brains largely have the same neurons but in different configurations).

Vertical Brain Integration

There is a series of practices that can be used to engage the prefrontal cortex and limbic circuitry to regain clear thinking and emotional regulation when stressors occur and cause distress or a stress reaction. Research demonstrates these 5 practices in ascending order of effectiveness to be significantly impactful in diminishing one's experience of emotional, psychological, and physical distress.

This occurs through intentionally slowing down and desiring to experience less distress as well as having increased GABA as a neurotransmitter that helps to decrease the impacts of cortisol and norepinephrine on the nervous system. As the neurotransmitters associated with the stress response subside, the prefrontal cortex becomes more "available" and the hippocampus can better regulate emotions. As a result, there can be a greater sense of calm and focus.

1. **Name it to tame it:** What emotion are you feeling right now?
2. **Control the environment:** In this situation, what can you change, do differently, or control? Can you leave the physical space that may be evoking the stress response?
3. **Values and life purpose:** What is important to you in this situation. What value do you want to honor?
4. **Reframing:** What is another way of looking at this situation, one you have not yet thought of? If you stood in someone else's shoes, what might this situation look like?
5. **Mindfulness:** Take a moment to breathe, connect with your senses and body. Draw 3–5 deep, full breaths and exhale slowly out of your mouth, using your diaphragm to push the breath all the way out. After you take this intentional pause, what are you now able to see about this situation in this moment?

Reflections

- How did naming the predominant emotion change your experience? Are there any other feeling "flavors" present?
- What strategies helped you regulate your stress response?
- How can you integrate this sequence into daily stress management?

Recovery, Renewal, and Restoration: The Power of Restorative Practices

The experience of being physically tired, emotionally depleted, and psychologically strained manifests to varying degrees and in individualistic ways for veterinary professionals at different points along their professional journey. Clarifying the various contributors to veterinary caregivers' experience of "fatigue" normalizes the conversation, offers suggestions from other caregiving professions that experience similar stressors, and allows individuals to better advocate for themselves. This increased awareness may then motivate those experiencing fatigue in their unique way to develop evidence-based skills and healthy coping strategies to better support life-work integration. The goal? To re-energize in the moment as needed and to stay connected to one's professional purpose and intentions.

To be clear, prioritizing quality rest and sleep is essential. Your body requires sleep, just as it needs air and food to optimize your health and overall functioning. During sleep, your body heals itself and restores its chemical and energy balance. Research found that after 17 hours without sleep, our alertness is similar to the effects of a blood alcohol concentration of 0.05%, which according to U.S. law is considered "impaired" on the legally drunk scale. After 24 hours without sleeping, the body is in the same state it'd be in at a BAC of 0.1% – well beyond the 0.08% BAC that identifies someone as legally drunk.[23]

Chronic sleep deficits are even more impactful than the one-off stint of long hours. The impacts can dramatically lower your quality of life, and research has demonstrated that sleeping too little per night (or period of rest if you are a shift worker) increases the risk of early death. These are serious enough consequences to make understanding this topic worth exploring and understanding further.[24]

Box Out: The Benefits of Healthy Sleep and Good Sleep Hygiene

Improved cognitive function: Having adequate sleep per cycle and enough time in both REM and Deep Sleep, supports the brain in forging new thought connections and helps memory retention – learning new information and technical skills.

Improved physical health and longevity: Adequate sleep promotes healing, restoration, and homeostasis of cardiovascular, musculoskeletal, CNS, and endocrine systems. And all of these measures may add years to your life as a result!

Improved mental health: Sleep plays a critical role in emotional regulation, decreasing anxiety and depression, and enhancing overall mental wellbeing.

Reduced risk of burnout: Increases emotional resilience, reduces circulating cortisol, increases stress management capacity, and contributes to healthy energy levels to meet the demands of veterinary work.

Promotes safety: Healthy sleep fortifies the crucial communications between neurons compromising the brain's signaling network. The lapses in memory and problems concentrating with chronic fatigue could result in self-injury, medical mistakes, and impairment overall in functioning in the world. Daytime drowsiness and microsleep episodes are less likely to occur . . . like while you are driving home after a long shift.

Sleep alone is not effective in managing exhaustion and fatigue. **Recovery, Renewal, and Restoration** are the pillars of sustainable wellbeing. In the demanding field of veterinary medicine, these practices are essential for maintaining mental, physical, and emotional health. **Recovery** involves intentional rest, **Renewal** is the act of revitalizing energy and purpose, and **Restoration** is the process of returning to a state of balance and wellbeing.

Why It Matters: Veterinary professionals often operate under high stress and emotional strain. Without intentional recovery and renewal, the risk of burnout and compassion fatigue increases dramatically. Implementing restorative practices helps maintain emotional resilience, improves cognitive function, and sustains energy levels.

Practices

- **Intentional Rest Periods:** Scheduling breaks intentionally throughout the day.
- **Digital Detox:** Designated tech-free times to reset mental focus.
- **Nature Exposure:** Time in nature to promote grounding and stress reduction.
- **Active Recovery Techniques:** Stretching, yoga, and light movement.
- **Sleep Hygiene:** Prioritizing consistent sleep schedules and healthy habits (blue light exposure, what/when you eat before bedtime, comfortable environmental temps).
- **Non-sleep Deep Rest:** Non-sleep deep rest (NSDR) is very much like yoga nidra practice and has impacts on brain waves.
  - By deactivating the sympathetic and activating the parasympathetic nervous system, NSDR may promote feelings of relaxation by increasing dopamine in the brain, while reducing heart rate and blood pressure.

- NSDR has been shown to promote brain rejuvenation, leading to improved memory consolidation and increased creativity. It is characterized by a reduction in beta waves, associated with active thinking and problem-solving, and an increase in alpha waves, indicating a state of relaxation.
- As the body enters this calmer state, brain waves decrease in frequency, thoughts slow, and stress hormones decrease.[25]

Strategies

- Set boundaries around work hours to protect recovery time.
- Implement a nightly wind-down routine to enhance sleep quality.
- Practice micro-restoration: 5-minute breaks for mindful breathing or stretching.
- Create a "Renewal Ritual": a practice that brings you joy and calm daily (mine is having a snack of dark chocolate!).
- Use weekend resets: Designate tech-free and/or nature-focused time blocks.
- Practice gratitude journaling as a form of emotional restoration.
- Leverage community-based recovery – group walks, yoga sessions, or debriefs.
- Establish non-negotiable boundaries for rest after intense work periods.

Phrases to Use

- "Rest is a form of productivity."
- "I am giving myself permission to restore."
- "I prioritize recovery so I can serve with strength."
- "Renewal is part of my self-care practice."
- "I protect my wellbeing through intentional rest."

Considerations

- Avoid sacrificing rest for productivity – long-term health depends on balance.
- Recognize signs of exhaustion early; recovery must be proactive, not reactive.
- Avoid the belief that rest is unproductive; it is crucial for resilience.
- Be mindful of tech overuse – digital breaks enhance mental clarity.

- Protect your time for restoration as fiercely as you protect work commitments.

Reflections

- What is one habit you can change today to prioritize restoration?
- Who can support you in maintaining boundaries around rest and renewal?
- How can you integrate more nature-based recovery into your routine?

Workbook Prompts

- Reflect on a time when you felt fully restored and energized. What contributed to that feeling?
- How do you currently prioritize rest and renewal in your routine? What could improve?
- Identify one area of your life where you can introduce micro-restoration practices.

Box Out: 7 Levels of Rest for Mind and Body

1. *Physical rest* (sleeping, napping, deep relaxation, yoga nidra, receiving a massage).
2. *Mental rest* (utilize "thought diffusion" to create objective distance between thoughts, emotions, and self; meditation; journaling before bedtime).
3. *Sensory rest* (turn off tech for designated amounts of time, e.g., 45 minutes before bedtime; grounding in your 5 senses; spend quiet moments outdoors).
4. *Emotional rest* (being authentic with how you are feeling with someone that you trust; name and accept the emotion you are experiencing and the impacts it has on you – positive and negative; practice self-compassion).
5. *Social rest* (know your "social battery" limits and know when to recharge; connect with a trusted person in your life where you feel safe and cared for; enjoy moments with your pets).
6. *Creative rest* (give your "doing" brain a break and move into your "being" brain to allow for inspiration, flow, and problem-solving to arise and inform your creative task; play music to soothe and/or inspire you; step outdoors and into feeling and broadened perspective).
7. *Spiritual rest* (connecting with something bigger than you through meditation, prayer, immersing yourself in something purposeful such as volunteering, being in "awe" with nature, music, or art).[26]

Connection and Community: Building Support Systems That Last

Connection and community are the foundations of resilience in demanding professions. In veterinary medicine, strong professional and personal support networks are vital for navigating emotional strain, ethical challenges, and day-to-day pressures. Building lasting connections fosters shared learning, emotional support, and collective resilience.

Why It Matters: Veterinary professionals often face isolation due to the intensity and emotional weight of their roles. Meaningful connections with colleagues, mentors, and community members help counteract feelings of isolation, provide emotional validation, and offer perspective during challenging times. Strong networks serve as buffers against burnout and enhance collective problem-solving.

Putting it into Action

Practices

- **Peer Mentorship:** Establishing regular check-ins with a mentor or peer.
- **Professional Learning Communities:** Engaging in collaborative learning with colleagues.
- **Collective Case Debriefs:** Reflecting on challenging cases with your team.
- **Support Circles:** Forming small groups for emotional processing and shared learning.
- **Community Engagement:** Volunteering or participating in community-based veterinary events.

Strategies

- Create a "Connection Plan" to intentionally build and maintain professional networks.
- Engage in regular peer debriefs to process emotional experiences collectively.
- Identify a mentor or accountability partner for guidance and support.
- Join professional associations or learning communities to widen your network.
- Organize team wellness activities to foster community.
- Use shared learning experiences as opportunities for connection and growth.

- Schedule monthly "Community Check-ins" with trusted colleagues.
- Participate in local or virtual support groups focused on veterinary wellbeing.

Phrases to Use

- "I value our connection and the support we share."
- "Let's debrief this case together."
- "I appreciate having a community that understands."
- "We're stronger together."
- "Support is part of sustainable practice."

Considerations

- Avoid isolating during times of stress – reach out, even when it's difficult.
- Be mindful of overcommitting to community roles; balance is key.
- Avoid relationships that drain your energy; seek those that renew and restore.
- Recognize that authentic connection takes time and intention.
- Avoid viewing networking as purely transactional – focus on genuine relationships

Reflections

- Who in your professional life can you lean on for support this week?
- What steps can you take to build a stronger sense of community?
- How would enhancing your professional network impact your resilience?

Workbook Prompts

- Reflect on a time when a colleague's support helped you through a challenging situation. What made that support impactful?
- How intentional are you about building professional relationships? What could improve?
- Identify one community or peer group that could enhance your support system.

Section 3: Sustainability Plan: Creating Holistic Personal and Professional Wellness

A sustainability plan is a personal project that helps identify and implement practices that support wellbeing across different aspects of life. It is a flexible, adaptable tool to align actions with values, needs, and aspirations. A well-structured sustainability plan fosters resilience, wellbeing, and satisfaction both personally and professionally. It is a living document, one that evolves with your growth, challenges, and changing needs. It is meant to be both a blueprint and a reflective practice that ensures you are living and working in alignment with your most deeply held values and goals.

Understanding the Essential Elements for Caregiver Flourishing

Building a sustainable plan for wellbeing is not a one-time effort; it is an ongoing practice of reflection, adaptation, and growth. The following sections break down the core elements that contribute to a holistic and thriving professional and personal life. Each segment includes actionable steps, reflective practices, and considerations to help you maintain alignment with your values and aspirations as you grow.

Emotional Intelligence: Self-Awareness and Self-Regulation/Mastery

Emotional intelligence (EI) refers to the ability to recognize, understand, and manage our own emotions while also recognizing, understanding, and influencing the emotions of others. In veterinary practice, where emotional intensity is often high, the ability to maintain self-regulation and clarity in emotionally charged situations is critical. EI is a key factor in effective communication, conflict resolution, and leadership. It equips us to navigate stress, empathize with clients, and manage the emotional weight of caregiving roles.

Why It Matters: Veterinary professionals often operate in environments charged with emotional intensity – clients worried for their pets, team members managing high workloads, and the constant presence of life-and-death decisions. Emotional intelligence serves as the anchor during these storms. By understanding and managing your own emotional state, you become more resilient, grounded, and capable of offering compassionate care without depleting your own wellbeing. EI also enhances team dynamics, as emotionally intelligent professionals can communicate more effectively, resolve conflicts with empathy, and contribute to a psychologically safe work environment.

Suggested Approach

- **Practices**: Journaling emotional triggers, mindfulness, active listening.
- **Strategies**: Name and tame emotions, seek feedback, practice empathy.
- **Phrases to Use**: "I notice that I'm feeling . . ."; "Help me understand your perspective."
- **Considerations**: Avoid emotional suppression, recognize burnout signs.
- **Reflections**: How did I manage my emotions today? What could I improve?

Workbook Prompts

- Reflect on a recent emotionally charged situation. How did you respond? What might you do differently next time?
- What are your top three emotional triggers, and how do you typically manage them?
- Who in your professional life demonstrates strong emotional intelligence? What can you learn from them?

Values Clarification and Application

Values are the deeply held principles that guide your decisions and influence your actions. Clarifying your values allows you to make choices that are consistent with your beliefs and aspirations. In caregiving professions, aligning work with your core values not only enhances job satisfaction but also fortifies you against moral distress and burnout. Knowing your values also serves as a compass during times of conflict or ethical dilemmas, enabling you to act with integrity and clarity.

Why It Matters: In the fast-paced and often emotionally demanding world of veterinary care, it is easy to be pulled in multiple directions that may conflict with your personal principles. When your values are clear, decision-making becomes more straightforward and resilient. Understanding what you stand for provides clarity in chaotic moments and strengthens your ability to set boundaries that protect your wellbeing. This values alignment is essential for maintaining satisfaction and purpose in your career.

Suggested Approach

- **Practices**: Values inventory, reflective journaling, decision audits.
- **Strategies**: Align choices with core values, recognize value conflicts early.
- **Phrases to Use**: "This aligns with my value of . . ."; "I prioritize [value] when making decisions."
- **Considerations**: Avoid compromising values for convenience.
- **Reflections**: Are my daily actions reflecting my core values?

Workbook Prompts

- List your top five values. How do these show up in your daily work?
- Recall a recent decision that felt misaligned with your values. What would you change?
- Identify an area of your professional life where your values are strongly represented. How does this alignment impact your wellbeing?

Positive Psychology and PERMA (Flourishing and Thriving)

Positive psychology shifts the focus from merely surviving to actively thriving. It emphasizes the strengths and virtues that enable individuals and communities to flourish. The PERMA model, developed by Dr. Martin Seligman, is a framework within positive psychology that highlights five core components of wellbeing: Positive Emotion, Engagement, Relationships, Meaning, and Accomplishment. In caregiving roles, these elements provide a pathway to sustained joy, resilience, and purpose, even amidst professional challenges.[27]

Why It Matters: In the field of veterinary medicine, where compassion fatigue and occupational distress are prevalent, positive psychology provides a science-based approach to enhancing resilience and job satisfaction. Focusing on what is going well, building strong relationships, and finding meaning in daily tasks can transform stress into growth opportunities. By intentionally cultivating the elements of PERMA, veterinary professionals can thrive – not just survive – in their careers.

Suggested Approach

- **Practices**: Gratitude journaling, strength-spotting, mindfulness.
- **Strategies**: Set achievable goals in each PERMA category.
- **Phrases to Use**: "What went well today?"; "How can I engage more deeply?"

- **Considerations**: Avoid the trap of constant achievement without reflection.
- **Reflections**: Which PERMA domain needs more attention?

Workbook Prompts

- List one meaningful accomplishment from the past week. How did it impact your sense of purpose?
- What relationships at work bring you the most joy and support? How can you nurture these more intentionally?
- Reflect on an area of your work where you feel deeply engaged. What conditions support that engagement?

Healthy Boundaries: Essential Self-Care

Boundaries are the personal limits we establish to protect our wellbeing, maintain energy, and foster healthy relationships. For veterinary professionals, clear boundaries are crucial for avoiding compassion fatigue and burnout. They serve as safeguards against overextending emotionally or physically, ensuring that your capacity to care for others is not compromised by self-neglect.

Why It Matters: In high-demand caregiving settings, the ability to create and maintain healthy boundaries is vital. Without them, the risk of empathic strain and burnout increases significantly. Healthy boundaries enable you to separate work from personal life, prevent emotional depletion, and create space for self-renewal. They are an act of self-respect that allows you to give your best without sacrificing your own health and happiness.

Suggested Approach

- **Practices**: Boundary-setting exercises, reflective self-checks, assertiveness training.
- **Strategies**: Communicate boundaries clearly, recognize overcommitment.
- **Phrases to Use**: "I'm not available at that time"; "I need some space to recharge."
- **Considerations**: Avoid saying "yes" out of guilt or obligation. If resentment shows up if you say a begrudging "yes," you may want to request a conversation to explore what a "qualified yes" or a full-on "no" might look like.
- **Reflections**: What boundaries need reinforcement in my life?

Workbook Prompts

- Identify one area where you need stronger boundaries. What would setting that boundary look like?
- How do you currently communicate your boundaries to others? Is it effective?
- Reflect on a time when you felt overwhelmed. Could setting a clearer boundary have helped?

Building Forward with Reflection and Adaptation

Each of these essential elements contributes to your ability to not only endure the challenges of caregiving but to flourish within them. By embracing emotional intelligence, aligning with core values, and actively participating in job crafting and professional development, you build a sustainable career that supports both personal and professional wellbeing.

The next step in this journey is understanding how to **Evaluate and Continuously Improve** your sustainability plan, ensuring it remains resilient and reflective of your evolving needs and goals.

Professional Development with Life in Mind

Professional development with life in mind is the intentional cultivation of career growth that is deeply intertwined with personal wellbeing and life goals. It means not only advancing your skills and knowledge but doing so with a clear understanding of your values, desired life path, and personal aspirations. This approach prevents the drift into unfulfilling roles and strengthens long-term career satisfaction.

Why It Matters: Veterinary professionals often face long hours, high emotional demands, and the risk of burnout. By intentionally aligning career growth with life goals, it is possible to craft a path that is both sustainable and fulfilling. This integration allows for more meaningful work experiences, better life balance, and resilience during professional challenges.

Suggested Approach

- **Practices:** Vision mapping, mentorship, reflective career audits.
- **Strategies:** Job crafting, long-term goal setting, pivot planning.

- **Phrases to Use:** "I am working toward . . ."; "This aligns with my career vision."
- **Considerations:** Avoid reactive career moves that are misaligned with long-term goals.
- **Reflections:** Is my career path sustainable for the life I want?

Workbook Prompts

- What are three career milestones you want to achieve in the next five years? Why do they matter to you?
- Reflect on your current role. How does it align with your long-term professional goals?
- List one professional skill you would like to develop further. What steps can you take to start?

Job Crafting to Avoid Bore-Out: Supporting Professional Stretching and Pivots

Job crafting is the proactive design and customization of your role to enhance personal engagement, fulfillment, and alignment with strengths and values. This concept recognizes that while job descriptions may be fixed, the way you experience and engage with your work is not. Through job crafting, you can reshape your tasks, relationships, and perspective to better fit your passions and skills, preventing stagnation and fostering deeper satisfaction.[28]

Why It Matters: In veterinary medicine, repetitive tasks, emotional strain, and limited role flexibility can lead to disengagement and what is termed "bore-out." Unlike burnout, which is driven by overwhelming demands, bore-out is the result of lack of challenge and purpose. Job crafting empowers you to re-engage with your work by adjusting tasks to be more meaningful, building stronger connections with colleagues, and reframing how you see your contributions. This proactive approach revitalizes purpose and helps sustain professional joy and commitment.

Suggested Approach

- **Practices:** Task redesign, relational crafting, cognitive reframing.
- **Strategies:** Engage in meaningful projects, seek learning opportunities, identify areas for innovation.
- **Phrases to Use:** "How can I reshape this role to be more engaging?"
- **Considerations:** Avoid passive acceptance of uninspiring tasks.
- **Reflections:** What small changes can I make to feel more engaged?

Workbook Prompts

- Reflect on an aspect of your current role that feels disengaging. What small change could make it more meaningful?
- Identify three strengths you bring to your role. How can you use these strengths to enhance your tasks?
- Describe one relationship at work that you would like to strengthen. What steps could you take to initiate that?

Courageous Professional Stretches

Courageous professional stretches are intentional efforts to step outside of your comfort zone in pursuit of growth, resilience, and mastery. These stretches can take many forms: assuming leadership roles, taking on challenging cases, public speaking, or deepening skills in a specialized area. The purpose of these stretches is to expand your capacity, build confidence, and explore new dimensions of your professional identity.

Why It Matters: Growth in any professional field – especially one as demanding as veterinary medicine – requires intentional risk-taking. Without deliberate challenges, skills can stagnate, and confidence may wane. Courageous stretches help to break the cycle of routine, prevent disengagement, and foster resilience. Stepping into unfamiliar territory, even when it feels uncomfortable, is where the most profound growth occurs. By seeking out opportunities to stretch, you actively contribute to your own professional development and enhance your sense of fulfillment.

Suggested Approach

- **Practices:** Volunteer for leadership opportunities, participate in professional panels, seek out case complexities that challenge you.
- **Strategies:** Set stretch goals, find a mentor for support, practice self-reflection.
- **Phrases to Use:** "I'm ready to step into this challenge"; "I want to explore this opportunity."
- **Considerations:** Avoid overextension, recognize your limits, and pace your growth.
- **Reflections:** What professional stretch would elevate my skills and confidence?

Workbook Prompts

- What is one professional challenge you have been avoiding? What is holding you back?
- Reflect on a time when you stepped out of your comfort zone. What did you learn from the experience?
- Identify a stretch opportunity within your current role. What support would you need to pursue it?

Support Systems and Peer Engagement

Support systems are the networks of relationships – both personal and professional – that provide encouragement, perspective, and resilience during challenges. Peer engagement includes intentional connection with colleagues who understand the unique demands of caregiving professions. In veterinary medicine, where emotional intensity is high, having a robust support system can be the difference between resilience and burnout.

Trusted colleagues, formal and informal mentors, and quality/healthy relationships outside of work are essential for an individual's holistic wellness. Particularly in times of growth and challenge, having meaningful support, psychological safety, specific feedback, and encouragement contribute to an individual's capacity and resilience.

Why It Matters: Navigating the complexities of veterinary practice is not meant to be done in isolation. Peer support helps mitigate feelings of isolation, validates shared experiences, and provides emotional grounding. Effective support systems create safe spaces for debriefing, shared learning, and collective resilience. Engaging with mentors, colleagues, and professional communities strengthens coping mechanisms and enhances professional satisfaction.

Suggested Approach

- **Practices**: Identify key support figures, join peer networks, participate in community groups.
- **Strategies:** Regular check-ins, peer mentoring, open communication.
- **Phrases to Use:** "I value your support"; "How can I support you?"
- **Considerations:** Avoid isolation – actively reach out during difficult periods.
- **Reflections:** Who can I lean on for support, and who can I support in return?

Workbook Prompts

- List three people in your professional network who provide support. How often do you connect with them?
- Reflect on a time when peer support helped you through a challenge. What made it impactful?
- Identify one way you can be a support to a colleague this week.

Mentorship and Peer Support

Definition: Building a support network of mentors and peers to enhance personal and professional growth.

Why It Matters: Meaningful, well-structured mentorship helps navigate career challenges and emotional complexities in caregiving environments.

Suggested Approach

- **Practices:** Identify potential mentors, set up regular check-ins, create peer support groups.
- **Strategies:** Foster reciprocal support, model vulnerability and growth.
- **Phrases to Use:** "I appreciate your guidance"; "I would value your perspective."
- **Considerations:** Avoid passive relationships; actively seek growth and learning.
- **Reflections:** How can I deepen my mentorship relationships?

Coaching and Professional Guidance

Coaching is intended to work in the present moment with someone as they identify a hurdle that they wish to overcome. Where therapy may examine the past to understand how someone got to where they are today, coaching (and professional mentorship) attend to the individual's current state of being and knowledge. Coaching, like mentorship, is based on trust, clear expectations, healthy boundaries, and specified goals. Coaching is different from mentorship in that the focus stays largely on the coachee's ability to get "unstuck" and move forward through the use of powerful questions and the development of new strategies and skills. The role of coach is to provide a space of supportive, informed accountability, not to give advice or share personal experiences. There are a wide variety of coaching styles, training/certification programs, and reasons why someone may choose a particular coach to navigate career and personal growth.

Why It Matters: Coaching provides structured support, accountability, and perspective during critical growth phases.

Suggested Approach

- **Practices:** Engage in regular coaching sessions, set clear goals, practice reflective dialogue.
- **Strategies**: Embrace feedback, pursue continuous learning, set SMART goals
- **Phrases to Use:** "I'm seeking guidance on . . ."; "What strategies would you recommend?"
- **Considerations:** Avoid stagnation – ensure progress is tracked and reviewed.
- **Reflections:** How has coaching influenced my path and decisions?

Evaluation and Continuous Improvement of the Sustainability Plan

In the creation of and "tweaking" of your unique "sustainability plan," you need to enter the GPS coordinates for the journey. Only you can know where you are and where you wish to head next.

Definition: Evaluation and continuous improvement are integral parts of sustaining a meaningful and resilient sustainability plan. This step involves regular reflection, assessment, and adjustment to ensure your plan remains relevant and effective as you grow and your circumstances change.

Why It Matters: A sustainability plan is not static; it should evolve as you do. Regular evaluation helps identify what is working well, what barriers are emerging, and where new opportunities for growth exist. It is a practice of mindful attention to your wellbeing and professional development, ensuring that your goals remain aligned with your core values and aspirations. When we intentionally reflect and adapt, we create space for greater resilience and a stronger sense of fulfillment.

Suggested Approach

- **Practices**: Schedule quarterly check-ins to review progress and obstacles.
- **Strategies**: Use reflective journaling, gather feedback from trusted peers, and track goal achievements.
- **Phrases to Use**: "What's working well right now?"; "Where am I feeling stuck?"

- **Considerations**: Avoid rigidity; embrace adaptability and be open to evolving needs.
- **Reflections**: How has my plan adapted to my changing needs? What would I like to improve?

Workbook Prompts

- Reflect on the past three months. What changes have you noticed in your professional or personal wellbeing?
- Identify one goal that you are struggling with. What adjustments could you make to improve your progress?
- List one new opportunity for growth that has surfaced recently. How can you integrate this into your sustainability plan?

Bringing It All Together: Your Personalized Sustainability Plan

To bring these concepts into actionable reality, the final step is to craft your *personalized* **Sustainability Plan**. The template here is designed to guide you through reflection, assessment, planning, and adaptation as you move forward.

Template for Building Your Own Sustainability Plan

Objective: To create a personal, actionable, and reflective sustainability plan for holistic wellbeing. This section guides you step by step through a structured approach to design and maintain your own plan, aligned with your values, aspirations, and professional goals.

Step 1: Reflection

This first step invites you to deeply consider your current state of wellbeing across different areas of your life. It is an opportunity to pause, assess, and set the foundation for purposeful growth.

Reflection Questions

- What areas of your life need more attention (emotional, physical, social, professional)?
- What are your core values and how do they align with your current actions?
- How satisfied are you in key areas: career, relationships, health, self-growth?

Workbook Prompts

- List three immediate changes you want to make to support your wellbeing. How can you begin this week?
- Reflect on your core values. Are there areas of your professional life where they are not well represented?
- Identify one professional habit that you would like to change or improve.

Step 2: Assessment

This step involves auditing your current state with tools like the Wellness Wheel (SAMHSA Wellness Wheel - Eight Dimensions of Wellness) at https://library.samhsa.gov/product/learn-eight-dimensions-wellness-poster/sma16-4953 or PERMA survey to identify strengths and gaps in each domain of life. This is where the realities of your wellbeing become clear and measurable.

Practices

- Perform a self-audit to map out where you feel aligned and where adjustments are needed.
- Use specific tools to measure aspects of wellbeing and professional satisfaction.
- Identify areas of growth and strength to build upon.

Step 3: Action Planning

With clarity from the assessment, you can now set actionable goals that are specific, measurable, achievable, relevant, and time-bound (SMART). This is where your vision becomes a structured pathway.

Practices

- Set SMART goals for each identified area of need.
- Outline 3–5 practices and 5–8 strategies to implement.
- Break down goals into actionable steps that fit into your daily routine.

Step 4: Implementation and Monitoring

This phase is about executing your plan and staying accountable. It is about moving from planning to action and measuring progress along the way.

Practices

- Create a timeline for each goal and designate checkpoints.
- Reflect monthly to assess progress and recalibrate if necessary.
- Seek accountability from mentors, colleagues, or self-check-ins.

Workbook Prompts

- What barriers did I face this month? How did I overcome them?
- What small wins can I celebrate today?
- Where do I need more support moving forward?

Step 5: Evaluation and Adaptation

Sustainability is not about static goals – it is about evolving with change. This step emphasizes the importance of regular review and adaptation to keep your plan relevant and aligned with your life.

Practices

- Review accomplishments and challenges quarterly.
- Adjust strategies, add new goals, or refocus based on current needs.
- Reflect on how well your sustainability plan is serving your holistic wellbeing and needs.

Reflection Questions

- What progress have I made?
- What barriers have I encountered, and how did I address them?
- How can I further align my actions with my values and aspirations?

Box Out: WOOP Method: Scientifically Proven Approach to Achieve Your Goals

- **Wish**: Think about a specific goal or wish that is challenging but doable.
- **Outcome:** Specify and imagine the best possible outcome. Imagine you have achieved your goal and think of all of the benefits you have gained.
- **Obstacles**: Identify the obstacles that could get in your way as you strive to achieve your goal.
- **Plan:** Create an "if-then" plan to overcome the obstacles and achieve your goal.[29]

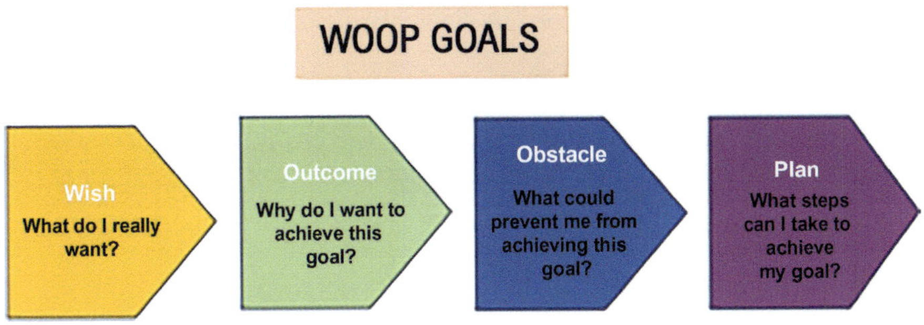

Figure 2.7 The WOOP method.

This 4-step process combines effective positive thinking with reality-acknowledging "mental contrasting" (the if-then plan) and with implementation intentions. This approach increases motivation and improves decision-making. By practicing the WOOP approach to reach your goals, you can cultivate an empowering mindset that also embraces flexibility and adaptability. This is a powerful and evidence-based tool to help you achieve your aspirations and optimize your full potential.

From Self to System: Why Individual Wellness Isn't Enough

You've just explored tools that help you **reset, recharge, and regulate your own wellbeing** – mindfulness, emotional agility, neurologic wisdom, and strategies to "stress better." These practices are powerful. They strengthen your resilience and capacity to thrive even when the work feels heavy.

But here's the truth: **no matter how skilled you become at caring for yourself, you do not work in a vacuum**. Veterinary medicine is a team sport. Your wellbeing and effectiveness are deeply intertwined with the health of the people around you – your colleagues, technicians, assistants, and leadership.

Think about it:

- What happens when you bring your best self to a team that is overwhelmed, disconnected, and running on empty?
- How sustainable is personal resilience when the surrounding culture fuels chronic stress, poor communication, or lack of trust?

The research is clear: **individual wellness cannot thrive long-term without team wellness**. Psychological safety, constructive communication, shared

Figure 2.8 Our Wellness Journey Continues from "Me" to "Us" in Chapter 3.

responsibility, and supportive systems are what transform coping into thriving.

Chapter 3 will take us there. We'll explore **what it means to create a culture of care**, where your tools for self-regulation meet strategies for **collective resilience and collaboration**.

This is the shift from "I" to "We" – and it's where the true magic of sustainable veterinary practice happens.

Note

1 Using the Emotion Wheel or your own prior experience, get curious and label the predominant emotion(s) present in this moment to bring clarity, as well as possible coping strategies, to mind.

References

1. Lyubomirsky, S. (2007). *The How of Happiness: A New Approach to Getting the Life You Want*. London: Penguin Books.
2. Seligman, M. E. P., Steen, T. A., Park, N., Peterson, C. (2005). Positive Psychology Progress: Empirical Validation of Interventions. American Psychologist, *60*(5), 410–421. https://doi.org/10.1037/0003-066X.60.5.410
3. Kabat-Zinn, J. (2005). *Wherever You Go, There You Are: Mindfulness Meditation in Everyday Life*. New York: Balance.
4. Kabat-Zinn, J. (2013). *Full Catastrophe Living: Using the Wisdom of Your Body and Mind to Face Stress, Pain, and Illness*. New York: Bantam.
5. Brier, J. N. (2015). Pain and Suffering. A Synthesis of Buddhist and Western Approaches to Trauma. In V. M. Follette, J. Briere, D. Rozelle, J. W. Hopper, D. I. Rome (Eds.), *Mindfulness-Oriented Interventions for Trauma: Integrating Contemplative Practice* (pp. 11–30). New York: The Guildford Press.
6. Brier, J. N., Scott, C. (2013). *Principles of Trauma Therapy: A Guide to Symptoms, Evaluation, and Treatment*. Newbury Park: SAGE Publications.
7. Woodward, L. (2024). *Mental Wellbeing and Positive Psychology for Veterinary Professionals: A Pre-Emptive, Proactive, and Solution-Based Approach*. Hoboken: Wiley.
8. Neff, K. *Self-Compassion Break*. Accessed May 2025. https://www.self-compassion.org/exercises/exercise-2-self-compassion-break
9. Brach, T. (2020). *Guided Meditation - The Practice of RAIN*. https://www.tarabrach.com
10. Hanson, R. (2013). *Hardwiring Happiness: The New Brain Science of Contentment, Calm, and Confidence*. New York: Harmony.
11. Flower, J. (1999). In the Mush. *Physician Executive*, 25(1), 64–66. PMID 10387273.
12. Brown, J. (2012). *The Art and Spirit of Leadership*. Bloomington: Trafford.
13. Dweck, C. (2017). *Mindset- Updated Edition: Changing the Way You Think to Fulfil Your Potential*. New York: Little Brown.
14. Pico, I. (2016). *The Wheel of Emotions, by Robert Plutchik*. https://psicopico.com.en/la-rueda-las-emocione-robert-plutchik/
15. Siegel, D. (1999). *The Developing Mind: How Relationships and the Brain Interact to Shape Who We Are*. New York City: Guildford Press.
16. David, S. (2016). *Emotional Agility: Get Unstuck, Embrace Change, and Thrive in Work and Life*. New York: Avery Publishing.

17. Reilly, R. (2024). *Completing the Stress Cycle: The Key to True Healing*. https://cptsdfoundation.org/2024/07/08/completing-the-stress-cycle-the-key-to-true-healing
18. Nagoski, E., Nagoski, A. (2019). *Burnout: The Secret to Unlocking the Stress Cycle*. New York: Ballantine.
19. Nash, W. P., Westphal, R. J., Watson, P., Litz, B. T. (2010). *Combat and Operational Stress First-Aid: Caregiver Training Manual*. Washington, DC: U.S. Navy, Bureau of Medicine and Surgery. Accessed October 4, 2020. https://www.academia.edu/20978265/Combat_Operational_Stress_First_Aid_Manual
20. Porges, S. (2011). *The Polyvagal Theory: Neurophysiological Foundations of Emotions, Attachment, Communication, and Self-Regulation*. New York City: W.W. Norton.
21. Pollatos, O., Kirsch, W., Schandry, R. (2005). On the Relationship between Interoceptive Awareness Emotional Experience, and Brain Processes. *Cognitive Brain Research*, 25, 948–962. https://doi.org/10.1016/j.cogbrainres.2005.09.019 CDC on sleep deprivation.
22. Barrett, L. F. (2020). *7 ½ Lessons About the Brain*. Boston: Harper Collins.
23. National Institute for Occupational Safety and Health, CDC. (2020). *Sleep Deprivation*. https://archive.cdc.gov/www_cdc_gov/niosh/emres/longhourstraining/sleepdeprivation.html
24. Gardiner, A., Quain, A. (2025). *The Veterinary General Practice Casebook: Companion Animal Clinics*. London: CRC Press.
25. Neuhaus, M. (2024). *Non-Sleep Deep Rest (NSDR): Exploring a World Beyond Sleep*. https://positivepsychology.com/non-sleep-deep-rest-nsdr/
26. Dalton-Smith, D. (2019). *Sacred Rest: Recover Your Life, Renew Your Energy, Restore Your Sanity*. Brentwood: FaithWords.
27. Seligman, M. E. P. (2011). *Flourish: A Visionary New Understanding of Happiness and Well-Being*. New York: Simon Element.
28. Wrzesniewski, A., Dutton, J. E. (2001). Crafting a Job: Revisioning Employees as Active Crafters of Their Work. Academy of Management Review, 26, 179–201. https://doi.org/10.1111/hsc.12084
29. Oettingen, G., Gollwitzer, P. M. (2018). Health Behavior Change by Self-Regulation of Goal Pursuit: Mental Contrasting with Implementation Intentions. In D. de Ridder, M. Adriaane, K. Fujita (Eds.), *The Routledge International Handbook of Self-Control in Health and Well-Being* (pp. 418–430). Routledge/Taylor & Francis Group.

3

Building Thriving Teams

Veterinary Medicine Is a "Team Sport"

The best protection against care giving providers being impaired by the innate emotional volatility and work-related psychosocial stressors is to provide opportunities to grow confidence and competence. This cannot occur in a vacuum nor in a work environment that does not prioritize the wellbeing of their very human resources. Individuals and teams are more likely to thrive when effective communication, constructive conflict management, and intentional preparation and ongoing training of non-clinical, essential skills are a priority for an organization and leadership.

Awareness of empathic strain, secondary traumatic stress, and moral conflict in veterinary teams is higher than it has ever been. That said, concrete strategies to use when these challenges arise at work for both self-regulation and for team support are essential to move forward with less energy-depletion and distress. Having increased peer support and leadership commitment to the team's wellbeing fortifies holistic wellness for the caregivers and career-sustaining behaviors. A more regulated, resourced, and energized veterinary team leads to greater professional satisfaction, better quality patient care, and values-aligned ways of practicing and communicating.

Let's lean into learning more about some of the most evidence-based elements of successful, interconnected teams. The next level of thriving for you is not a solo endeavor.

This chapter moves us from personal mastery to collective mastery, from "I" to "We."

DOI: 10.1201/9781003475361-3

Intentional Communication: Building Connection and Trust

What Is Intentional, Effective Communication?

Intentional communication is the purposeful use of verbal and nonverbal skills to foster mutual understanding, trust, and psychological safety in relationships. It involves self-awareness, emotional regulation, and a deliberate focus on forming authentic connections, particularly important in high-stress, emotionally charged environments like veterinary medicine.

Effective communication is not just about transmitting information; it's about being understood and fostering a relationship. It supports shared decision-making, demonstrates empathy, and allows for emotional expression without judgment.

"What can I do to build this relationship?" becomes the guiding mindset.

Why It Matters in Veterinary Wellbeing

Human relationships are complex, and we are emotional beings by nature. Nonverbal cues like tone, gestures, posture, and facial expressions account for 80% of communication and affect trust. A study on communication between veterinarians and clients revealed that the clients were interrupted typically within 15–30 seconds of starting to speak.[1] This is likely not unique to veterinary environments or to the conversations between veterinary clinicians and clients. What is more probable, as you likely know from your life experience, is that people often tend to listen to reply rather than understand. We are in our heads thinking about what to say, perhaps how to be right, or about how to move the conversation along and be efficient if we are "busy." There are strategies to change this habitual behavior and instead practice curiosity, humility, and patience. Competent, intentional communication builds:

- Trust and rapport
- Emotional resilience (for team and client)
- Collaborative decision-making
- Better outcomes for patients and professionals

Relationship-centered communication isn't just a "nice to have" – it's core to wellbeing and sustainable practice.

Core Practices and Strategies for Intentional Communication

The Four Foundational Skills

1. **Empathy**
   - Acknowledge and normalize emotions.

   - Use sincere, caring statements:
     - "I hear how much this matters to you."
     - "That sounds incredibly difficult."
2. **Open-Ended Inquiry**
   - Invites dialogue and reflection:
     - "What's been most challenging for you?"
     - "How can I best support you right now?"
3. **Active Listening ("listening to learn")**
   - Presence is power. Use body language and silence wisely.
     - Make eye contact.
     - Pause before responding.
     - Avoid distractions.
4. **Reflective Listening**
   - Paraphrase to clarify:
     - "So what I'm hearing is . . ."
     - "Did I understand you correctly when you said . . .?"

Mindful Nonverbal Communication

- Pay attention to facial expressions, tone, posture, proximity.
- Use calm, steady vocal tone and open body language.
- Be mindful of cultural, emotional, and situational nuances.

Additional Techniques

- **Pausing**: Center yourself before speaking.
- **Chunk and Check**: Offer info in digestible parts; ask for understanding.
- **Permission Language**:
  - "Would it be okay if I share a thought?"
- **Signposting**:
  - "First we'll talk about the options, then we can discuss what feels best to you."
- **Tailor Your Words**: Match the other person's knowledge and emotional capacity.

What to Avoid

- Diminishing others' experiences:
  - Don't say, "I know how you feel."
- Inauthentic empathy:
  - Avoid scripted or insincere comments.
- Assuming emotional states:
  - Replace "You seem angry" with "I'm noticing some tension – how are you doing?"

Workbook Section

Self-Reflection Prompts

- When do I communicate with the most intention?
- How do I react when conversations become emotionally charged?
- What are my go-to nonverbal cues – and how might they be received?

Practice Scenarios

- **Reformat closed questions** into open-ended ones.
- **Write an empathetic response** to the following client concern:

"I can't believe this is happening. I don't know what to do."

Phrase Bank for Relationship-Centered Conversations

Scenario	Intentional Phrase
Client is overwhelmed	"It's completely understandable to feel that way."
Team member is struggling	"I'm here if you want to talk through it."
Clarifying instructions	"Just to make sure we're on the same page, what I heard was . . ."

Worksheet Activity: My Communication Inventory

- What are three verbal habits I want to improve?
- What are two nonverbal patterns I want to be more aware of?
- What's one strength I bring to difficult conversations?

Closing Insight: Communication Is Daily Practice

> *"The practice is the practice."*

Intentional communication isn't about perfection, it's about progress. When we communicate with clarity, curiosity, and compassion, we become not only better caregivers but also more resilient, connected humans.

Trauma-Informed Care in Veterinary Environments

What Is Trauma-Informed Care?

Trauma-informed care (TIC) is an approach that acknowledges the widespread impact of trauma, recognizes its signs and symptoms, and responds by fully integrating knowledge about trauma into policies, procedures, and practices.[2] In veterinary settings, TIC fosters environments of psychological safety, trust, and compassion.

A trauma-informed mindset shifts from asking "What is wrong with you?" to "What is happening with you?" – cultivating empathy, connection, and mutual respect.

Six guiding principles anchor trauma-informed systems:

1. Safety (especially psychological)
2. Trustworthiness and transparency
3. Peer support
4. Collaboration and mutuality
5. Empowerment, voice, and choice
6. Cultural, historical, and gender sensitivity

Why Trauma-Informed Care Matters

Veterinary medicine is emotionally demanding, with inherent exposure to high stress, grief, and moral distress. *Secondary traumatic stress* and *empathic strain* wellbeing can erode (please see Glossary for definitions of these terms).[3] When not recognized and attended to with healthy coping strategies by individuals and teams, there can be significant and detrimental impacts, particularly over time, on physical and mental health, relationships, and professional efficacy. Research on caregiving professionals, including in veterinary medicine, has shown that these psychosocial concerns combined with moral conflict are significant contributors to occupational burnout.

When the guiding principles of a trauma-informed approach are part of hiring, training, and culture creation for a team and practice, the following are promoted:

- Resilience and recovery in individuals and teams
- Emotional intelligence and psychological safety in workplaces
- Reduction in burnout, moral injury, and compassion fatigue

- Strengthened trust between clients, caregivers, and leadership
- Cultures where empathy, support, and learning are prioritized

Positive Impacts of Trauma-Informed Care

For Clients

- Builds trust and communication
- Reduces feelings of judgment or shame
- Encourages openness and cooperation

For Veterinary Staff

- Normalizes emotional experiences
- Encourages self-awareness and skill building
- Enhances professional fulfillment and meaning

For Organizational Culture

- Supports transparency and collaboration
- Promotes healthy peer and leadership relationships
- Encourages sustainable care environments

Practical Strategies and Workbook Activities

Daily Practices

- Pause and reflect before reacting: "How can I meet this moment with compassion?"
- Practice emotional regulation through breathing or grounding techniques
- Ask team members regularly: "What support would feel most helpful right now?"

Trauma-Informed Workplace Behaviors

- Normalize the human emotional experience
- Offer content/experience warnings before difficult cases
- Promote clarity and honesty: avoid false reassurances like "Your pet will be fine"
- Prioritize empowerment: include all voices in decision-making when feasible

Peer and Leadership Strategies

- Build debriefing and peer-support systems

- Ask open, supportive onboarding questions:
  - "What helps you feel supported when work feels hard?"
  - "What should I know to help you succeed here?"
- Use decision-making models (e.g., IDEA) for ethical challenges

Box Out: IDEA Ethical Decision-Making Framework

This is one of many approaches available, and in general the veterinary profession is still "young" in developing our understanding of bioethics and of ethical literacy. The IDEA Ethical Decision-Making Framework was developed by the Regional Ethics Program based at Trillium Health Partners and incorporates principles from various ethical frameworks.

1. **Identify** the facts (gather relevant information about the situation and the people involved)
   Ask: What is the ethical issue?
2. **Determine** the Relevant Ethical Principles (understanding personal and societal values and ethical principles relevant to the situation; what will the impacts be of different decisions on all stakeholders?)
   Ask: Have perspectives of relevant individuals been sought?
3. **Explore** the options (harms/benefits; strengths/limitations; laws/regs: mission/vision/values; generating and evaluating alternatives)
   Ask: What is the most ethically justifiable option?
4. **Act** (recommend the best course of action consistent with ethical principles and values; implement;, evaluate, and maintain an open, honest dialogue throughout the process)
   Ask: Are we (am I) comfortable with this decision?[4]

Recommended reading on ethical issues and management in veterinary medicine:
Veterinary Ethics: Navigating Tough Cases (2017) by Siobhan Multan and Anne Fawcett

Reflections

- What does psychological safety mean to me?
- How can I support myself and others after a challenging event?
- What signs tell me I'm approaching burnout or secondary trauma?

Worksheet Activity: Trauma-Informed Self-Awareness Inventory

- What are my stress warning signs?
- What calming or coping strategies work best for me?
- What trauma-informed behaviors do I want to strengthen in my team?

Closing Thought: Compassion Is a Practice

Adopting a trauma-informed approach is not about fixing others, it's about creating a culture of awareness, responsiveness, and resilience. Through understanding, shared responsibility, and daily kindness, we foster veterinary environments where all beings – human and animal – can thrive.

Connection and Community: Building Support Systems That Last

Connection and community are the foundations of resilience in demanding professions. In veterinary medicine, strong professional and personal support networks are vital for navigating emotional strain, ethical challenges, and day-to-day pressures. Building lasting connections fosters shared learning, emotional support, and collective resilience.

Why It Matters: Veterinary professionals often face isolation due to the intensity and emotional weight of their roles. Meaningful connections with colleagues, mentors, and community members help counteract feelings of isolation, provide emotional validation, and offer perspective during challenging times. Strong networks serve as buffers against burnout and enhance collective problem-solving.

Putting it into Action

Practices

- **Peer Mentorship:** Establishing regular check-ins with a mentor or peer.
- **Professional Learning Communities:** Engaging in collaborative learning with colleagues.
- **Collective Case Debriefs:** Reflecting on challenging cases with your team.
- **Support Circles:** Forming small groups for emotional processing and shared learning.
- **Community Engagement:** Volunteering or participating in community-based veterinary events as a team/practice

Strategies

- Create a "Connection Plan" to intentionally build and maintain professional networks.
- Engage in regular peer debriefs to process emotional experiences collectively.
- Identify a mentor or accountability partner for guidance and support.
- Join professional associations or learning communities to widen your network.
- Organize team wellness activities to foster community.
- Use shared learning experiences as opportunities for connection and growth.
- Schedule monthly "Community Check-Ins" with trusted colleagues.
- Participate in local or virtual support groups focused on veterinary wellbeing.

Phrases to Use

- "I value our connection and the support we share."
- "Let's debrief this case together."
- "I appreciate having a community that understands."
- "We're stronger together."
- "Support is part of sustainable practice."

Considerations

- Avoid isolation during stress – reach out to trusted others, even when it's difficult.
- Be mindful of overcommitting to community roles; balance is key.
- Avoid relationships that drain your energy; seek those that renew and restore.
- Recognize that authentic connection takes time and intention.
- Avoid viewing networking as purely transactional – focus on genuine relationships.

Reflections

- Who in your professional life can you lean on for support this week?
- What steps can you take to build a stronger sense of community?
- How would enhancing your professional network impact your resilience?

Workbook Prompts

- Reflect on a time when a colleague's support helped you through a challenging situation. What made that support impactful?
- How intentional are you about building professional relationships? What could improve?
- Identify one community or peer group that could enhance your support system.

Stress First Aid (SFA)/Psychological First Aid

We explored Stress First Aid a bit in the last chapter as it applied to individual recognition and coping strategies for stress reactions when they occur. Similar to physical stress, psychological stress can cause injury to the mind and body. These injuries go beyond burnout: a **stress injury** is any severe and persistent distress or loss of ability to function caused by damage to the brain, mind, or spirit after exposure to the overwhelming stressors of fatigue (burnout), trauma, loss, or moral injury.[5] In particular, there is a significant risk of moral injury in high-stress, service-oriented professions where valued qualities such as selflessness, loyalty, a strong moral code, and excellence can also create vulnerabilities, such as prioritizing the needs of others above one's own needs. Veterinary medicine is certainly one of these professions, so veterinary professionals would benefit from knowing more about the factors that contribute to stress reactions and how to navigate them with more awareness and skill.

The SFA training space and toolbox building includes insightful and practical approaches to support teams through compassionate approaches to and through stress reactions as they occur for co-workers and for the entire team.

Stress First Aid (SFA) is a peer-support model designed to help individuals and teams recognize, address, and recover from stress reactions. It is based on the Stress Continuum Model (see Figure 3.1), which classifies stress responses from "Ready" (Green) to "Reacting" (Yellow), "Injured" (Orange), and "Ill" (Red).[5]

Why This Matters

"Why" a structured, evidence-based, and proactive approach like SFA matters for veterinary teams is similar as to why it would matter for individual workers.

The Stress Continuum Model

READY (Green)	REACTING (Yellow)	INJURED (Orange)	ILL (Red)
DEFINITION □ Optimal functioning □ Adaptive growth □ Wellness FEATURES □ At one's best □ Well-trained and prepared □ In control □ Physically, mentally, and spiritually fit □ Mission-focused □ Motivated □ Calm and steady □ Having fun □ Behaving ethically	DEFINITION □ Mild and transient distress or impairment □ Always goes away □ Low risk CAUSES □ Any stressor FEATURES □ Feeling irritable, anxious, or down □ Loss of motivation □ Loss of focus □ Difficulty sleeping □ Muscle tension or other physical changes □ Not having fun	DEFINITION □ More severe and persistent distress or impairment □ Leaves a scar □ Higher risk CAUSES □ Life threat □ Loss □ Moral injury □ Wear and tear FEATURES □ Loss of control □ Panic, rage, or depression □ No longer feeling like normal self □ Excessive guilt, shame, or blame □ Misconduct	DEFINITION □ Clinical mental disorder □ Unhealed stress injury causing life impairment TYPES □ PTSD □ Depression □ Anxiety □ Substance abuse FEATURES □ Symptoms persist and worsen over time □ Severe distress or social or occupational impairment

Figure 3.1 The Stress Continuum Model.
Source: Adapted and reproduced with permission from Richard J. Westphal. This work was prepared at the request of the US Government. Title 17 USC 105 provides that "Copyright protection under this title is not available for any work of the United States Government." Title 17 USC 101 defines a United States Government work as work prepared by a military service member, employee, or contractor of the United States Government as part of that person's official duties.

Just as I said in Chapter 2, veterinary professionals often face high-pressure environments that trigger acute and chronic stress. SFA provides a structured method to support oneself and others, promoting recovery before stress reactions escalate into larger problems that may impact health, relationships, or career stability.

Practices

- **Recognize the Signs**
  - Identify symptoms early (e.g., emotional reactivity, withdrawal, agitation).
- **Provide Support**
  - Validate experiences and listen actively.
  - Offer resources and practical support when needed.
- **Encourage Recovery**
  - Promote self-care practices and peer support.
  - Create an environment that fosters psychological safety.

Strategies

- **The Stress Continuum Model**: Assessing where someone is on the spectrum of stress.
- **Early Intervention:** Addressing yellow zone reactions before they escalate.
- **Peer Support Training:** Teaching team members to recognize and respond supportively. (Many practices are creating *Wellness Ambassadors* or *Champions* for their teams).
- **Creating Safe Spaces:** Establishing environments where emotions can be shared openly.
- **Routine Check-Ins:** Regularly assessing emotional wellbeing among staff.

Phrases to Use

- "I'm here to support you, how can I help?"
- "I see you're going through a lot. Let's talk it through."
- "I recognize that this situation is really challenging. How can I best support you?"

Considerations

- Avoid minimizing someone's stress reaction.
- Be aware of compassion fatigue in peer-support roles.
- Ensure that interventions are voluntary and consensual.

Reflections

- Consider a recent stressful event – what phase of the "stress continuum" were you in?
- Who supported you during that time, and what was helpful?
- How might you extend SFA principles to colleagues or team members?

Critical Incident Group Debriefing and Huddles – Taking a Multidisciplinary Approach

Another strategy that is huge in building a multidisciplinary team (MDT) approach is group debriefing. In a group debrief, everyone with a perspective on a particular event, patient, or situation is asked to come together. This is not a "fact-finding" gathering; instead, it is a moment to intentionally suspend social judgment and listen to one another answer a formatted series of questions:

1. What is your viewpoint on this event? This is not a recalling of facts, but instead, what keeps you awake at night thinking about this, or what are the most relevant thoughts that are repeating in your mind?
2. Where did you see your individual strengths show up? What about your teams' strengths?
3. What were the biggest challenges for you and your team?
4. Where do you see this growth, personally or professionally?
5. What are you most grateful for? These questions promote rich perspective sharing rooted in genuine interest and trust.

The group debrief can be utilized in several spaces, such as in morbidity and mortality rounds or when discussing near-miss patient outcomes. In addition to exploring what can be learned, these conversations also have great value in underpinning the "miracle" cases and highlighting when things go well.

The "team huddle" is another instrument of an MDT approach. When teams face critical situations requiring the utmost expertise, having role-clarifying huddles before these events occur allows for transactive, muscle-memorized behaviors when needed. Huddles can also be called when a team member needs to lean into their team's collective intelligence to innovate a diverse array of solutions and provide the best possible care. Case reports are also a technique utilized to enhance collaboration. Bringing forth cases and inviting team input and feedback is a great way to show that everyone on the team has a vital role in cases and that each one has an influence in meeting the best outcomes.

Box Out: Co-Regulation of Stress Reactions

As veterinary teams, there is the opportunity to lean into compassionate 'co-regulation' when work-related stressors occur. Co-regulation refers to the process where individuals adapt their actions to regulate one another in a cooperative way.[7] If you and your team are not yet practicing low-impact debriefing following critical medical or patient-safety events, or after working through challenging client communications, get curious about the different approaches that exist around ethical and impactful debriefing. Facilitated debriefing sessions can reduce emotional overwhelm, foster team cohesion, and protect against long-term impacts like burnout, compassion fatigue, and PTSD.

There are specific training modules, webinars, and articles available on critical incident group debriefing (CIGD) strategies and protocols for leadership and team trainers to utilize in support of themselves and their staff. Additionally, there are veterinary community support spaces that can provide additional resources around these important and emotional conversations:

- Reviving Veterinary Medicine as a community has trained CIGD facilitators that offer support (https://revivingvetmed.com).
- Veterinary Hope Foundation has small group discussion spaces to support individuals through stressful professional experiences (https://veterinaryhope.org).
- VIN Foundation's Vets4Vets and Support4Support offers confidential peer support and a wide range of small group community discussions for veterinarians and team members (https://vinfoundation.org/resources/vets4vets).
- Vetlife is another valuable veterinary-centric resource hub supporting veterinarians, nurses, practice managers, and other non-clinical staff with 3 support services: a 24/7 helpline, Health Support, and Financial Support (https://vetlifeorg.uk/).

Healthy Boundaries for Sustainable Teams and Leadership

In Chapter 2, healthy boundaries for an individual were shared as essential self-care. In this chapter on supporting the creation and ongoing "maintenance" of healthy, resilient teams, the concept of boundaries is equally critical.

What Are Healthy Boundaries for Teams?

Healthy boundaries are limits we establish to protect our energy, time, emotional wellbeing, and values. In a veterinary team context, boundaries enable professionals to show up with integrity, clarity, and presence – without becoming overwhelmed or depleted.

These boundaries exist not only at the individual level but also at the cultural and organizational levels. When teams and leadership foster a space of psychological safety, mutual respect, and support for life-work integration, boundaries become a shared foundation for professional interconnectedness and sustainability.

Why They Matter

- Boundaries support self-care, energy management, and holistic wellness.

- They foster respect, autonomy, and clear communication.
- When respected, boundaries decrease emotional reactivity and improve trust.
- Boundary-rich cultures improve team morale, reduce burnout, and enhance retention.[8]

A workplace that honors healthy boundaries empowers individuals to take ownership of their wellbeing while cultivating a collaborative, accountable team dynamic.

Workbook Section

Practices/Action Steps

- **Define Your Boundary** – Identify what you need to feel respected and balanced.
- **Communicate Clearly** – Express needs simply and directly without over-explaining.
- **Use High-Brain Communication** – Set boundaries in a calm, regulated state.
- **Respect Others' Boundaries** – Practice healthy receivership.
- **Reinforce With Values** – Frame boundaries around what matters to you (e.g., "I value focused patient care, so I turn off notifications during procedures").

Strategies and Approaches

- **Create Personal Boundary Scripts** – Practice saying "no" kindly but firmly.
- **Model Boundary Setting as Leaders** – Normalize and support it throughout the team.
- **Use SOPs to Clarify Roles and Expectations** – Reduces ambiguity and protects time.
- **Support Flexible Scheduling Where Possible** – Encourages life-work harmony.
- **Encourage Micro-Breaks and PTO Use** – Allow for cognitive and emotional resets.
- **Promote Delegation** – Prevent over-functioning and promote growth.
- **Provide Communication Training** – Support assertive, respectful language.
- **Empower Teams to Problem-Solve** – Build autonomy within boundaries of respect.

Phrases to Use (Communication Tools)

- "I'm at capacity right now and can't take that on."
- "I need a few minutes to reset before we continue."
- "Here's what I'm available for today."
- "I respect that request, but I'll need to decline."
- "Let's revisit this conversation when I have more bandwidth."

Considerations (Barriers and Pitfalls)

- **Guilt** – Feeling selfish or rude for asserting needs.
- **Over-explaining** – Undermines clarity and confidence.
- **Inconsistent reinforcement** – Sends mixed signals.
- **Lack of leadership modeling** – Makes boundary-setting feel unsafe.
- **Boundary resistance from others** – Often stems from discomfort with change.

Reflections ("How Am I Doing?")

- What boundaries have I set recently that supported my wellbeing?
- Where do I need clearer limits or stronger follow-through?
- How do I typically respond when someone sets a boundary with me?
- How can I advocate for a more boundary-respecting culture in my team?
- What do I need (from myself, team, or leadership) to support life-work integration?

Closing Insight: Boundaries Are Acts of Respect

Boundaries aren't barriers, they are bridges to more intentional, empowered engagement. By cultivating a culture where personal limits are honored and communicated with care, veterinary teams can thrive together with integrity, sustainability, and mutual trust.

Constructive Conflict and Positive Conflict Resolution

What Is Constructive Conflict?

Constructive conflict is the process of engaging in open, respectful, and solution-focused discussions when disagreements arise. Rather than avoiding or escalating conflict, this approach recognizes it as a natural part of human interaction – and an opportunity for growth, deeper understanding, and relationship strengthening.[9]

This approach involves self-awareness, emotional intelligence, and a shared commitment to solving problems collaboratively. At its core, constructive conflict supports trust, innovation, and psychological safety.

Why It Matters

- **Conflict is inevitable** in dynamic, high-stress environments like veterinary medicine.
- **Unresolved conflict** contributes to team dysfunction, burnout, and toxic culture.
- **Constructive conflict** strengthens relationships, sharpens communication skills, and improves team performance.
- It **enhances mutual understanding and enables problem-solving** through diverse perspectives.

When managed with empathy and intention, conflict becomes a catalyst for connection and progress.

Workbook Section

Practices

- **Define the Core Problem** – Identify what the conflict is truly about before jumping to solutions.
- **Pause to Reflect** – Ground yourself emotionally before responding. Breathe, then engage.
- **Create a Safe Space** – Choose a time and place conducive to open, honest conversation.
- **Use "I" Statements** – Focus on expressing your experience rather than blaming.
- **Clarify a Shared Goal** – Anchor the discussion in common purpose (e.g., team harmony, patient care).

Strategies and Approaches

- **Active Listening** – Allow the other person to speak without interruption; reflect back what you hear.
- **Humble Curiosity** – Ask open-ended questions: "Can you help me understand your perspective?"
- **Acknowledge Emotions** – Validate what others feel, even if you see things differently.

- **Identify Conflict Styles** – Know your default (avoidant, competitive, accommodating, compromising, collaborating).
- **Focus on the Issue, Not the Person** – Keep the conversation solution-oriented.
- **Review and Reframe Options Together** – Explore alternatives and compromise when possible.
- **Debrief after Resolution** – What worked? What could we do differently next time?
- **Create a Conflict SOP** – Establish team norms for addressing conflict respectfully and promptly.

Box Out with Figure 3.2: **Conflict Styles – Based on Thomas Killman Conflict Modes Model**

1. Competing
2. Collaborating
3. Compromising
4. Avoiding
5. Accommodating[10]

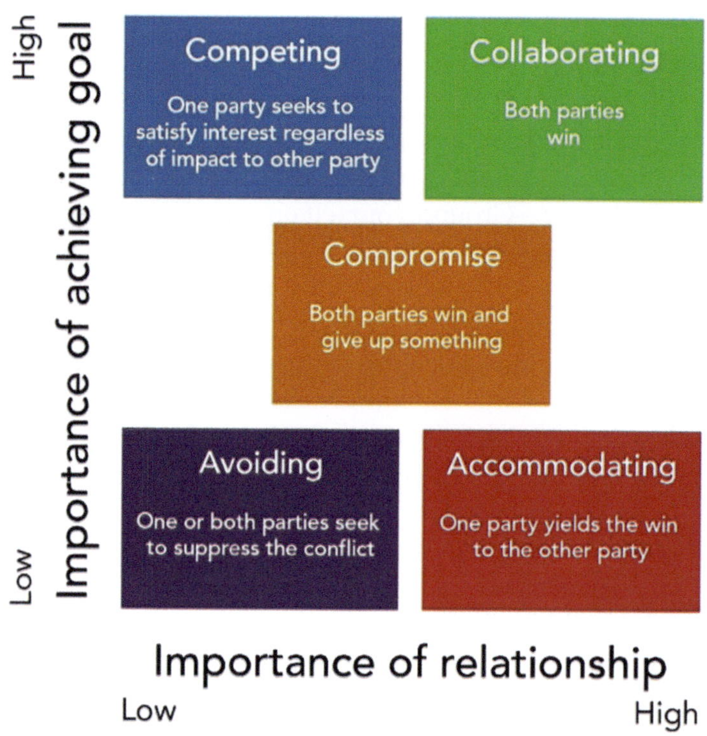

Figure 3.2 Thomas Kilman conflict styles.
Graphic credit: Olson, 2022, p. 128–9).

Most individuals have 1–2 styles that they rely upon and use them when appropriate in situations:

- Which to use and why?
- How important are the issues to you? To the other?
- How important is it to maintain a positive relationship?
- Is the issue time sensitive (e.g., a dog has been hit by a car and treatment needs to be determined immediately. . .)?
- To what extent does one party trust the other?
- What is the power differential in the conflictual relationship?[11]

Phrases to Use (Communication Tools)

- "I'm noticing some tension – can we talk about what's going on?"
- "I feel concerned about how that interaction went and would like to understand better."
- "What's most important to you in this situation?"
- "Let's work together to find a way forward."
- "Here's what I heard – did I get that right?"

Considerations (Barriers and Pitfalls)

- **Avoiding the issue** – Hoping it goes away often allows resentment to fester.
- **Blame language** – "You always . . ." or "You never . . ." fuels defensiveness.
- **Lack of follow-up** – Agreements without accountability often fall flat.
- **Ignoring nonverbal cues** – Body language often reveals underlying discomfort.
- **Fixing too quickly** – Jumping to solutions can shut down emotional processing.

Reflections ("How Am I Doing?")

- How do I typically respond to conflict – avoid, compete, accommodate, collaborate?
- What recent conflict did I handle well? What contributed to that?
- Where do I need to grow in communication under stress?

- What kind of environment helps me feel safe to share a different perspective?
- What support or training would help me feel more confident in navigating conflict?

Closing Insight: Conflict as a Bridge

Conflict does not have to divide us. When approached with courage, clarity, and compassion, it becomes a bridge – to stronger teams, better outcomes, and a more resilient, respectful veterinary workplace.

Team-Based Medicine: Strengthening Veterinary Care through Collaboration

As we collectively explore how to create the veterinary profession of tomorrow that benefits all stakeholders (caregivers, support staff, leaders, clients, veterinary patients, practice owners), team-based medicine and thoughtful utilization of the entire veterinary team's skills will be foundational pillars for success and sustainability. We have been moving in this direction for decades since the role of veterinary technician/registered animal nurse was first created in the United States, United Kingdom, Australia, and Canada in 1961.

Note: India and several countries in Latin America are now considering following the standards and training guidelines to formally create the job role of veterinary technician/nurse in the near future.

Team medicine is a model of veterinary practice that emphasizes optimal use of all team members – veterinarians, credentialed veterinary technicians/nurses (CrVTs/VNs), assistants, and client care staff – to deliver the best patient outcomes. It involves clear role definitions, trust-based collaboration, and opportunities for growth through cross-training and professional development. There are excellent resources that have been recently created by experienced veterinary professionals that inform this section of the workbook:

- The 2023 AAHA Technician Utilization Guidelines [12]
- Workplace Strategies to Reduce Burnout in Veterinary Nurses and Technicians (2025)[13]
- VetPartners 2024 Veterinary Team Utilization Guide[14]

Let's examine where we are today with awareness, research, and application of concepts from inside and outside of veterinary medicine to create practical and impactful strategies for veterinary team medicine globally.

Technician/Nurse Utilization

What It Is

The full and appropriate use of credentialed veterinary technicians (CrVTs) and veterinary nurses (VNs) to perform all tasks within their scope of practice.

Why It Matters

- Enhances efficiency and patient care quality.
- Increases job satisfaction, reducing attrition and burnout.
- Supports practice profitability, with studies showing ~$100,000+ additional annual revenue per veterinarian when technicians are fully utilized.
- Frees veterinarians to focus on diagnosis, surgery, and client communication.

Strategies and Approaches

- Delegate advanced tasks (anesthesia monitoring, client education, teletriage).
- Create SOPs and workflows that reflect all utilization levels (basic to advanced).
- Educate veterinarians and managers on technician capabilities and scope.
- Build trust through open dialogue on delegation and shared accountability.

Workbook Prompts

1. List three tasks you currently perform that could be delegated to a technician/nurse.
2. How might increased technician utilization improve both patient care and your workday flow?

Job Role Clarity

What It Is

Clear definition and communication of responsibilities, expectations, and accountability for each role on the veterinary team.

Why It Matters

- Reduces confusion, overlap, and conflict.
- Strengthens accountability and efficiency.
- Creates alignment with practice values and goals.
- Provides a foundation for trust and psychological safety.

Strategies and Approaches

- Develop and share written job descriptions that include technical and interpersonal skills.
- Align responsibilities with legal practice acts.
- Use leveling systems to clarify progression and expectations.
- Regularly review and update role clarity during staff meetings.

Workbook Prompts

1. Write down the top 5 responsibilities of your current role. How clearly are these responsibilities understood by your colleagues?
2. Where do you notice overlap or role confusion in your team?

Cross-Training

What It Is

Training team members to develop competency in multiple roles, allowing for flexibility and resilience.

Why It Matters

- Increases adaptability during staff shortages or emergencies.
- Enhances empathy and teamwork by helping team members understand each other's challenges.
- Prevents bottlenecks by ensuring multiple staff can cover critical tasks.

Strategies and Approaches

- Identify key skills across roles that are trainable (e.g., basic client communication, patient restraint, inventory management).
- Implement phased training plans to prevent overwhelm.
- Pair cross-training with mentorship for support.
- Rotate roles during non-critical times to practice new skills.

Workbook Prompts

1. Identify one skill from a colleague's role you would like to learn. How might this benefit the team?
2. What is one skill you could teach another team member to strengthen team resilience?

Professional Development

What It Is

Ongoing education, training, and career pathway opportunities that build skills and foster career satisfaction.

Why It Matters

- Enhances employee retention and engagement.
- Builds long-term resilience and career sustainability.
- Supports leadership growth within the team.
- Increases practice reputation and client trust through advanced skills.

Strategies and Approaches

- Offer structured mentorship programs for new graduates and staff.
- Provide continuing education (CE) opportunities and leadership training.
- Develop progression pathways (entry → advanced → leadership).
- Embed adult learning principles to support skill retention.

Workbook Prompts

1. What is one area of veterinary medicine or teamwork you would like to develop further?
2. How could your practice create or strengthen career progression opportunities?

Trust and Relationships

What It Is

The mutual respect, psychological safety, and collaborative spirit that allow veterinary teams to work together effectively.

Why It Matters

- Strong trust reduces conflict and builds team resilience.
- Improves communication, delegation, and morale.
- Directly protects against burnout by fostering a sense of belonging and shared purpose.

Strategies and Approaches

- Hold regular team huddles and debriefs.
- Encourage open, respectful communication where all voices are valued.

- Recognize and celebrate contributions across all roles.
- Train leaders to model trust-building behaviors.
- Refer back to Chapter 2's section on *Support Systems and Peer Engagement* for additional approaches to support the entire team's experience of safety, support, and growth.

Workbook Prompts

1. Think of a time when you felt deeply trusted in your veterinary role. What actions or systems supported that trust?
2. What is one step you could take this week to strengthen trust within your team?

Box Out: 6 Thinking Hats Technique

The **Six Thinking Hats technique**, developed by Edward de Bono, is a structured method for problem-solving and decision-making that involves six different styles of thinking, each represented by a colored hat:[15]

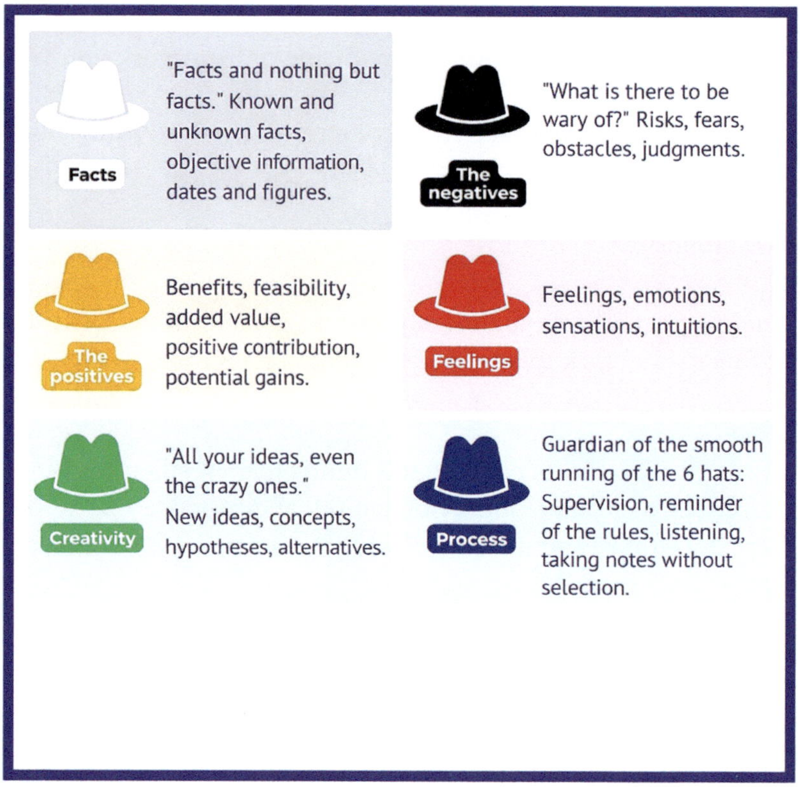

Figure 3.3 The Six Thinking Hats technique.
Source: Permission through Imfusio.com.

Integrating Boundaries, Conflict Resolution, and Trauma-Informed Care through Collaborative Team Agreements

Why Inclusive Team Agreements Matter

Collaboratively crafted team agreements act as living documents that align group norms with values of respect, trust, equity, and psychological safety. They offer a foundation where healthy boundaries, constructive conflict resolution, and trauma-informed principles can thrive together, supporting both individual wellbeing and team cohesiveness.

These agreements help clarify expectations, promote mutual accountability, and operationalize the six core principles of trauma-informed care:

1. **Safety** – Emotional and psychological safety for all team members
2. **Trustworthiness and Transparency** – Clarity in decision-making and communication
3. **Peer Support** – Encouraging collaborative problem-solving and mutual encouragement
4. **Collaboration and Mutuality** – Shared power and teamwork in addressing challenges
5. **Empowerment, Voice, and Choice** – Valuing diverse input and individual autonomy
6. **Cultural, Historical, and Gender Sensitivity** – Recognizing and honoring lived experiences

What a Collaborative Team Agreement Can Look Like

- A shared document co-created by all team members and reviewed regularly
- Includes clear, respectful communication norms
- Reflects a commitment to healthy boundary setting and mutual support
- Offers a common language for navigating conflict
- Anchored in values like empathy, equity, and wellness

Workbook Section: Crafting Collaborative Team Agreements

Practices/Action Steps

1. **Host a Values-Based Workshop** – Invite your team to explore personal and collective motivators.
2. **Co-Create the Agreement** – Ensure every voice is included in identifying guiding principles and ground rules.

3. **Assign a Maintenance Role** – Designate a rotating team member to revisit, review, and suggest updates.
4. **Incorporate the 6 TIC Pillars** – Use them as lenses for assessing team practices.
5. **Celebrate Wins** – Acknowledge moments where the agreement supports team health and cohesion.
6. **Collaborative Coaches** – Facilitate productive conversations and uphold the values and tenets of the team agreements.

Strategies and Approaches

1. **Draft Respectful Communication Norms** – Define how the team handles disagreements.
2. **Embed Boundary Language** – Include permission to say no and reset without guilt.
3. **Design Feedback Pathways** – Include safe, scheduled opportunities to offer team/process feedback.
4. **Integrate Conflict SOPs** – Outline how conflict will be navigated using a shared framework.
5. **Schedule Regular Debriefings** – Normalize check-ins after challenging days or experiences.
6. **Support Flexibility in Structure** – Allow team members to voice what works best for their life-work integration.
7. **Foster Psychological Safety** – Embed statements of shared intent like, "We assume positive intent and honor repair."
8. **Use Trauma-Informed Prompts** – For example, "What support do you need right now?" or "What's your capacity today?"

Phrases to Use (Communication Tools)

- "Can we revisit our agreement to see how it's supporting us?"
- "What does support look like for you today?"
- "Let's bring this to our next team norms review."
- "Is this a boundary we need to revisit together?"
- "How do we hold space for each other while honoring what we need?"

Considerations (Barriers and Pitfalls)

- **Top-Down Creation** – Agreements created without team involvement can feel performative.

- **Outdated or Unused Documents** – Agreements that are not revisited lose relevance.
- **Avoiding Hard Conversations** – Naming discomfort is essential for growth.
- **Uneven Power Dynamics** – Create mechanisms for every voice to be heard safely.
- **Rigid Norms** – Agreements should evolve with the team.

Reflections ("How Are We Doing?")

- What values are most important for our team to embody?
- Do we feel safe giving and receiving feedback?
- What's one team practice that helps me feel supported?
- What part of our agreement is working well? What needs to be adjusted?
- How can we better integrate trauma-informed, boundary-honoring, conflict-ready practices?

Closing Insight: Agreements Anchor Culture

A collaborative team agreement is more than a document – it's a mirror of how a team aspires to show up together. When rooted in compassion, accountability, and shared growth, these agreements transform culture from the inside out.

From "We" to the Workplace: Cultivating Environments Where Veterinary Teams Flourish

In the previous chapters, we have journeyed from individual wellbeing practices to the collective strength of teams. By developing awareness, resilience, and intentional strategies at the personal and group levels, we have built a strong foundation for veterinary professionals to thrive together. Chapter 3 highlighted that veterinary medicine is not a solo pursuit but a true "team sport," where communication, role clarity, psychological safety, and shared responsibility foster resilience and connection.

The next step in this wellness journey is to broaden our perspective to the practice environment and organizational levels. Sustainable wellbeing does not rest on individuals or teams alone – it requires systems, structures, and leadership that actively support and model a culture of health. Veterinary leaders hold a unique responsibility to shape environments where compassion, psychological safety, and professional growth are not only encouraged but expected. When leaders and organizations invest in these priorities, the benefits extend beyond the caregivers themselves: clients experience greater trust, practices achieve stronger cohesion, and patients receive the best possible care.

As we move into Chapter 4, we will explore leadership development and practice-level strategies that bring these principles to life. This includes cultivating compassionate and positive leadership, embedding psychological safety, supporting resilience, and designing organizational cultures that foster thriving for all. By embracing this shared responsibility, veterinary practices can become places where individuals, teams, and communities of care flourish together.

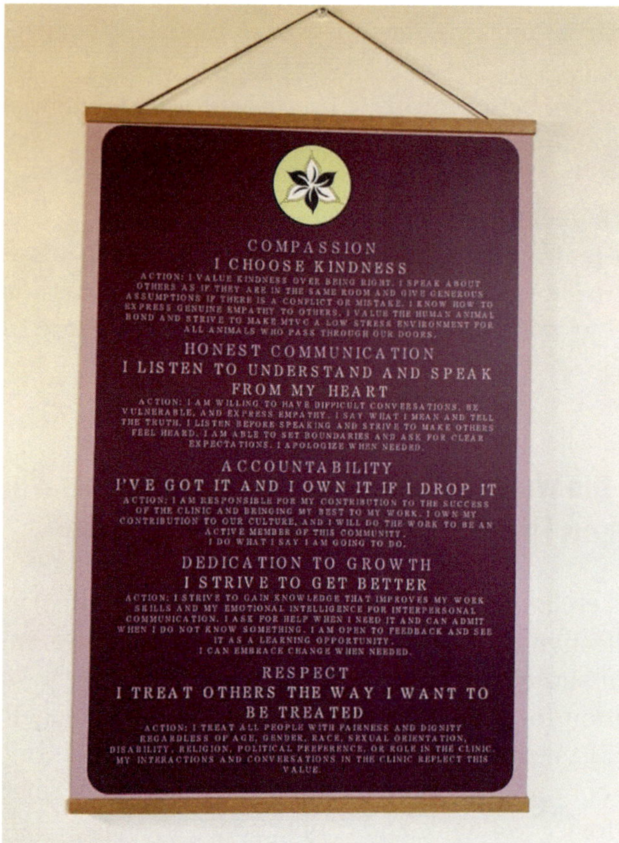

Figure 3.4 This Team Values Map illustrates the foundational values that guide healthy veterinary teams in their collaborative agreements – serving as a compass for respectful communication, shared accountability, and trauma-informed care.
Source: Photograph taken by author and with the permission to share by the owners of Mt. Tabor Veterinary Care, Portland, OR.

Figure 3.5 The Journey Continues: From Me to We to Us.

References

1. Janke N, Shaw JR, Coe JB. On-site communication skills education increases appointment-specific client satisfaction in four companion animal practices in Texas. *JAVMA*, 2022;260(13):1711–1720. https://doi.org/10.2460/javma.22.06.0242
2. Substance Abuse and Mental Health Services Administration. *SAMHSA's Concept of Trauma and Guidance for a Trauma-Informed Approach. HHS Publication No. (SMA) 14-4884*. Rockville, MD: Substance Abuse and Mental Health Services Administration, 2014.
3. Moir F, Van den Brink A. Current insights in veterinarians' psychological wellbeing. *New Zealand Veterinary Journal*, 2020;68(1):3–12. https://doi.org/10.1080/00480169.2019.1669504

4. Trillium Health Partners. *IDEA Ethical Decision-Making Framework Guide.* https://trilliumhealthpartners.ca/IDEA-Framework-THP
5. Shanafelt T, Stolz S, Springer J, Murphy D, Bohman B, Trockel M. A blueprint for organizational strategies to promote the well-being of health care professionals. *NEJM Catalyst*, 2020;1(6). https://doi.org/10.1056/CAT.20.0266
6. Nash WP, Westphal RJ, Watson P, Litz BT. *Combat and Operational Stress First-Aid: Caregiver Training Manual*. Washington, DC: U.S. Navy, Bureau of Medicine and Surgery, 2010. Accessed October 4, 2020. https://www.academia.edu/20978265/Combat_Operational_Stress_First_Aid_Manual
7. Welldoing.org. *Co-Regulation: How the People Around You Impact Your Nervous System*, 2024. https://welldoing.org/article/co-regulation-how-people-around-you-impact-your-nervous-system/
8. Tawwab NG. Set Boundaries Workbook: Practical Exercises for Understanding Your Needs and Setting Healthy Limits. Los Angeles: Tarcher-Perigee, 2021.
9. Kriesberg L. *Constructive Conflicts: From Escalation to Resolution*. Lanham: Rowman & Littlefield, 1998.
10. Thomas K, Kilmann R. The social desirability variable in organizational research: An alternative explanation for reported findings. *Academy of Management Journal*, 1975;18(4):741–752.
11. Olson S. *Creating Wellbeing and Building Resiliency in the Veterinary Profession: A Call to Life*. London: CRC Press, 2022.
12. AAHA. *AAHA Technician Utilization Guidelines*, 2023. https://www.aaha.org/resources/2023-aaha-technician-guidelines/
13. Chapman AJ, Bennett PC, Rohlf VI. Workplace strategies to reduce burnout in veterinary nurses and technicians: A delphi study. Animals, 2025;15:1257. https://doi.org/10.3390/ani15091257
14. VetPartners. *Veterinary Team Utilization Guide*, 2025. https://www.utilization-guide.vetpartners.org
15. De Bono E. *Six Thinking Hats*, Revised Ed. Toronto: Penguin, 2000.

4

Sustainably Healthy Practices

Leadership Development and Culture Cultivation

We live and work in a "VUCA" world. This term refers to the fact that the concerns that we navigate in our personal, professional, and societal realms are "Volatile, Uncertain, Complex, and Ambiguous." Specifically, the acronym is based on the leadership theories of Warren Bennis and Burn Nanus in the 1990s and has since spread to apply to strategic leadership in organizations when faced with challenges and opportunities. Leaders can use the system of thought as it has since evolved to see change not just as inevitable but something to anticipate.[1]

Creating environments of skillful communication and listening with the intentional development of a safety culture is essential to tackle the important and vital work of problem-solving needed today in many realms of our personal and professional lives. The veterinary industry is grappling with how to meet the challenges and opportunities of today's global dynamics inside and outside the profession in a sustainably healthy manner. This requires innovative, dedicated approaches and intentional actions grounded in multidisciplinary knowledge and skill utilization.

International research efforts over the last several decades into what elements contribute to healthy leaders, teams, and work culture have led to a cross-profession list. The following list from a recent leadership development and team-building article is reflective of current thinking.

DOI: 10.1201/9781003475361-4

Key Leadership Skills and Competencies

1. Emotional intelligence
2. Relationship building
3. Problem solving
4. Decision making
5. Persuasion
6. Coaching
7. Compromise
8. Delegation
9. Time management
10. Organization
11. Boundary-setting
12. Presentation skills
13. Active listening
14. Empathy
15. Effective serial tasking (humans cannot multitask!)
16. Focus
17. Learning
18. Discretion
19. Planning and strategizing
20. Emotional regulation
21. Instruction
22. Conflict resolution
23. Resilience[2]

In this chapter, we will examine some of these "buckets" of skill and wisdom that fortify sustainably healthy and prepared leaders, which is foundational to create a culture of wellness for the team and a thriving business of providing best care to animals.

The other skills listed can be found strengthening the leadership approaches explored in this chapter. In each section, there are practices, strategies, phrases, considerations, and workbook prompts offered to offer strength and confidence building opportunities in your work environment.

Section 1: Compassionate Leadership

Let's look at compassionate leadership as a foundational approach to effective, sustainable leadership.

Definition

Compassionate leadership integrates empathy with action. Empathy is the ability to recognize and feel another person's emotions; compassion moves beyond this to offer supportive, often practical action. It is empathy + action, rooted in human connection and the desire to alleviate suffering.

> **Box Out: Real-Life Scenario For Context**
>
> **Empathy:** You see a colleague upset after a patient euthanasia and you feel a pang of sadness as well.
>
> **Compassion:** You see the colleague experiencing sadness *and* you decide to check in with them "Hey, that was tough. Do you want to take a moment to talk? I'm here to listen."

Why It Matters

Managers and leaders are often overwhelmed and unequipped to lead change, with 69% feeling ill-prepared.[3] Compassionate support is essential, as it can mitigate distress, increase resilience, and enhance their ability to lead effectively. Networking opportunities and peer connections to share current trends and best practices can be more beneficial than traditional training, helping leaders embody and enforce cultural values and support their teams through change-fatigue.

Far from being just a *feel-good* concept, compassion is rooted in neurobiology. Research by Goetz, Keltner, and Simon-Thomas shows that compassion activates neural networks associated with caregiving and reward. It fosters emotional regulation, stress resilience, and improved team bonding.[4] In veterinary medicine, where teams face moral distress and loss, compassionate leadership prevents burnout and builds collective capacity to thrive.

Compassion as a Balanced Approach

Experts argue that it's not "compassion" that drains workers and leaders in caregiving environments but the prolonged or repetitive exposure to distress without boundaries or solutions – empathy that lacks the supportive action side. If we only absorb others' pain and suffering without a strategy to handle it, empathic strain can occur.[5]

> **Box Out: Neuroscience Insight**
>
> "True compassion includes boundaries. Without them, the very act of helping can lead to depletion or harm." [6]
>
> – Singer and Klimecki (2014)

Practices to Fortify Compassionate Leadership

- Check in with team members during tough moments.
- Model self-compassion and boundary-setting.
- Offer peer support after difficult procedures.
- Introduce mindfulness or gratitude rituals at team meetings.

Strategies

- Provide compassion-focused communication training.
- Establish a buddy system or emotional debriefing protocol.
- Recognize and celebrate moments of compassion satisfaction.
- Incorporate trauma-informed care principles into team interactions.

Phrases to Use

- "That was tough – how are you feeling?"
- "Thank you for being here, especially on a hard day."
- "What can I do to support you right now?"

Considerations

- Compassion without boundaries leads to fatigue and contributes to burnout for leaders.
- Normalize taking emotional resets.
- Compassion must be structurally supported.

Reflections

- When have I seen compassion shift a situation?
- How does my own self-compassion shape my leadership?

Workbook Prompts

- Describe a time a compassionate act changed your day.
- What is one compassionate practice you can start this week *for yourself?*

Key Takeaways

1. **Empathy** is feeling another's emotion; **compassion** adds a desire to help.
2. Our **brain** is wired with mirror neurons and reward systems that encourage caring.
3. Compassion reduces stress and fosters **emotional resilience**, helping combat depression, anxiety, and loneliness.
4. **Self-compassion** counteracts harsh self-criticism, fueling motivation and personal growth.
5. Regular **training** (like loving-kindness meditation or mindful listening) literally **rewires** neural pathways for more compassionate responses.
6. True compassion includes **boundaries**, preventing burnout and ensuring you can help others without losing yourself.
7. A culture of compassion – whether in the family, workplace, or community – can produce stronger social bonds and a more peaceful environment.

Section 2: Positive Leadership and the 4 Ps

Veterinary environments, particularly those serving emergency and critical care patients, are complex social systems with innately challenging emotions. Navigating a myriad of stressful circumstances individually and collectively is part of the job description. Creating awareness and a working knowledge of how to develop a culture of trauma-informed care, which includes psychological safety, will mitigate some of the negative impacts on physical and mental health for veterinary caregiving teams and those that support these individuals.

When considering "safety," the first thought may be of Maslow's hierarchy of needs, where safety is considered a basic human need.[7] You are likely to have a familiarity with the safety culture as it applies to physical safety for caregivers and the more structured reporting of medical mistakes. As a result, many veterinary practices have established safety standards and protocols. Likely the inclusion of the elements of psychological safety and trauma-informed care, as newer concepts to the veterinary industry, are not considered as part of protocols and ways of working, teaming, and learning.

Over the last several decades, human medicine and other professional work environments have demonstrated value in supporting the holistic health of individuals and of teams by fostering a culture that does take psychological safety and trauma-informed communication into consideration. As relatively unfamiliar concepts to many in the veterinary profession, it would be valuable to first define both of these terms. This sets the foundation for exploration of the potential positive impacts on individual associates, on teams, and on the larger veterinary systems that veterinary professionals work within.

Definition

Positive leadership is an intentional, strengths-based approach that fosters agency, hope, and engagement. It integrates emotional intelligence, character strengths, and systemic thinking to promote workplace wellbeing.[8]

Why It Matters

Positive leaders elevate team morale, performance, and innovation. In emotionally demanding fields like veterinary medicine, they buffer stress, increase resilience, and reinforce meaning in work.

Key Traits of Effective Positive Leaders

- Self-awareness (asking for feedback)
- Relational transparency (having a clear leadership philosophy)
- Balanced processing (using active listening)
- Ethical behavior (following through on what you say you will do)
- Trustworthiness (treating others with respect and keeping your word)
- Supportiveness (giving appreciation and support to followers)
- Empowerment (giving your followers freedom and choice)[9]

Practices to Support Positive Leadership

- Acknowledge effort and micro-wins, not just outcomes.
- Practice strength-based leadership by recognizing and celebrating the strengths of your team.
- Use inclusive language.
- Be transparent about challenges and re-frame them as opportunities ("we get to . . .").

Strategies

1. **Implement the 4 Ps**

Psychological Safety

- The belief that you won't be punished or humiliated for speaking up with ideas, questions, concerns, or mistakes.
- Supports an atmosphere of authenticity, growth, learning, and trust between team members.
- Increased confidence, creativity, and engagement occur when psychological safety is present in the workplace.
- Curiosity, speaking up, and a culture of honest, respectful communication and knowledge-sharing spurs innovation and improves quality of care and medicine; for example, discussion of mistakes is seen as a learning opportunity, not for blame/shame.

Purpose

- Routinely show that all work matters.
- Connect all work to a higher purpose.

Path

- Clear expectations of individuals, of job roles, of timelines, of collective values and goals.
- Autonomy and self-efficacy are enabled to support professional growth and mastery.
- Resources are available for individuals to succeed.

Partnership

- Individuals on the team feel seen, heard, appreciated and valued.
- Each team member plays a crucial role. Everyone becomes a vital part of the organization's mission and purpose and contributes to the collective knowledge that supports positive, effective change.

2. **Encourage gratitude rounds.** *At the end of the work shift, as part of patient rounds or passing on work responsibilities, intentionally create the opportunity to notice and express what went well, to appreciate a teammate, and to reconnect with collective purpose. All of these fortify interpersonal relationships and resiliency within teams.*

Phrases to Use

Team leads/shift supervisors/PMs may lead the discussion with these queries:

- "What is one thing that you feel went well today?"

- "I want to take a moment to appreciate (name) for (what they did/said)."
- "How did we make an animal, or family's, life better during our shift?"

Resource: The Science of Gratitude: How It Changes Your Mind[10] (https://www.sciencenewstoday.org/the-science-of-gratitude-how-it-changes-your-mind)

3. **Set up mentorship and collaboration opportunities**. All roles and levels of experience benefit from peer support and the opportunity to learn from one another. Cross-training is an excellent way to create an awareness and appreciation of what others on your team are experiencing and are responsible for in their role. This also creates flexibility, rapport, and increased professionalism in teams.

Phrases to Use

- "You are essential to our mission."
- "How are we progressing together?"
- "What are some of your professional goals? How can we support these goals? (*Note: consider using the SMART+Evaluated and Reviewed strategy of bringing goals to life in specific, time-boxed, action-oriented ways*).

Positive Leadership – Considerations

- Following through and timely communication is key to trust.
- Be authentic and avoid "toxic positivity," for example, buying pizza for the team when things get tough rather than addressing the questions/concerns of the staff in a safe, open way. Toxic positivity may show up as being "tone deaf" or not acknowledging that more conversation is needed for topics, plowing forward with an agenda despite there being a clear desire to slow down/pause and discuss certain topics, and it may be that the leader's own self-awareness around their choice of language, tone, body language say one thing while their words say another. All humans are exceptional at discerning when there is dissonance between what is being said and what the speaker actually feels/believes.

Reflection: How Do I Currently Lead with Purpose and Partnership?

Workbook Prompts

- Which of the 4 Ps needs more support in your workplace?
- What is one strategy (or phrase) you can implement right away to create impactful changes?

> **Box Out: From the Literature**
>
> "Positive affect and psychological safety are linked to better performance and job satisfaction."[11]
>
> – The National Standard of Canada for Psychological Health and Safety in the Workplace

Section 3: Emotional Intelligence and Psychological Safety

In both human and veterinary medicine care environments, emotional intelligence and psychological safety have been studied and proven crucial to:

- Foster collaboration, trust-building, and skillful communication, which are essential for highly effective and innovative teams
- Elevate the quality of medicine provided and thus patient safety and outcomes
- Support the holistic health of individual veterinary professionals, teams, and organizations[12]

Although we touched on emotional intelligence and psychological safety in Chapter 2, I am bringing them forward here again as essential elements for healthy, successful leaders and ecosystems of caregiving work. Some of the most successful and healthy businesses and teams in the world are those that have embraced being fluent in the language, practices, and applications of emotional intelligence and psychological safety. Resiliency, innovation, and improved professional health and satisfaction have been found to be fortified in workplaces that include emotional intelligence and psychological safety as essential leadership skills and as values for the entire team.

Definition

Emotional intelligence (EI) is the ability to understand and manage one's own emotions while navigating the emotions of others. Psychological safety is the belief that one can express thoughts and feelings without fear.

Why It Matters

EI enables deeper connection and stress regulation. Psychological safety drives learning, innovation, and resilience.

Emotional Intelligence

The term "emotional intelligence" was coined by Daniel Goleman, an American psychologist, author, and science journalist. For twelve years, he wrote for *The New York Times*, reporting on the brain and behavioral sciences. It was in his impactful 1995 book *Emotional Intelligence: Why It Matters More than IQ* that he first brought his research, contextual application, and effective strategies to the world.

Dr. Goleman's 5 Core competencies of EI

- Self-Awareness
- Self-Regulation
- Motivation
- Empathy
- Social Skills – Social Perception and Social Effectiveness[13]

Today, although there is no authoritative definition of EI, all definitions have internal (personal) and external (social) dimensions.

Self-awareness is the foundation of cultivating improved EI in all regards. Increasing and maintaining a higher degree of self-awareness supports better self-control, also called self-management. In the fortification of the skills in the internal (personal) dimensions, there is increased agency and ability to be both more present in your life and in control of how you show up in it. Living and behaving in a values-aligned, intentional way serves you and the community that you care about it and brings increased focus and energy to the endeavors that matter most in your life, both personal and work-related.

Practice: Self-Awareness Cultivation

- Keep checking within yourself. Utilizing a growth mindset and embracing that learning and living mean constant change. This is a wonderful opportunity to practice being in the present moment and releasing overthinking or identification with our thoughts.
- Practice mindfulness for self-awareness – these go hand in hand. Exist with your thoughts and emotions without judgment, as they do not define you. They are signals and data to guide you toward deeper self-knowing.

- Keep a journal. Record and process feelings and experiences to learn and grow with more intention and values-aligned behavior rather than spinning in anxiety or rumination.
- Complete a self-awareness test (there are so many online options!). How much do you understand yourself? What are your strengths, and what are opportunities for improvement? You could also do this in real time: Evaluate yourself in a particular situation and then ask colleagues and/or staff members to share how they might evaluate you in the same situation. This is like a mini-360 assessment but focused on a specific situation in which you may be keen to build confidence and competence.
- Be more of a listener. Introspection and deep listening to learn are catalysts on the journey of self-awareness. When in conversation, also consider whether what you said with good intentions landed the way you had thought that it would. Ask and hear what is shared. Own that what you said may have caused confusion or harm, and retry with increased awareness, intention, and clarity.
- Be open to and accept feedback from those individuals with whom you have wholesome relationships. This provides the opportunity to gain insights into how others perceive our character as well as how different individuals have various perspectives of us.

Box Out: Quote Offered by Dr. Adam Grant to Challenge Perceptions

"Ask people for feedback, and they sometimes tell you what you want to hear. Ask them about blind spots, and they're more likely to tell you what you need to hear. Gains in self-awareness often begin with the question: What do other people know about me that I may not realize?"[14]

When self-awareness and self-management skills are developed, the external EI skills can be more available. This creates the conditions for an increased awareness of the impact our behaviors have on other people, of their emotions, and of nonverbal communication such as body language. Relationship-building is fortified by asking clarifying questions, showing empathy, and utilizing healthy boundary-making and conflict management strategies.

Overall, when EI is high, there is the greater ability to live a life aligned with our values and integrity and to positively influence others.

Why EI Matters

Your brain is wired to react emotionally first in the limbic region long before the signals travel to the reasoning, rational part of the prefrontal cortex.

Good news! The combination of the neuroplasticity of your brain and the positive changes that can occur with intentional repetitions of "noticing" and modifying behavior can result in the healthy rewiring of your brain, supporting increased emotional intelligence.

Dr. Travis Bradberry, a world-renowned researcher and author on emotional intelligence, writes in his 2025 "The New Emotional Intelligence":

> The motivation behind every action is inherently emotional. Communication between the rational and emotional centers of your brain is the essence of emotional intelligence. Your rational thoughts, driven by an understanding of your emotions, give you the opportunity to channel your emotions into producing the behavior that you want.[15]

Dr. Bradberry offers his latest EI assessment tool that can be taken twice to gauge progress with the skill-building efforts associated with this book. His model of EI focuses on 4 EI Skills: Self-Awareness, Self-Management, Social Awareness, and Relationship Management.

The book provides 60 in-depth strategy-building opportunities in each of these skill areas that are worth exploring. These are some additional and impactful practices to get the ball rolling.

Practices

Radical Acceptance: Release the "what ifs" and focus on the present moment and "what is." This is leading from a place of emotionally regulated response and realism rather than from fear.

Practice Mindful Check-Ins: Pause throughout the day every few hours and ask, "What am I feeling right now?"

"Name It to Tame It": Dr. Dan Siegel's "Name It to Tame It" method is based on the idea that when we put our feelings into words, we activate the brain's self-regulation systems. This technique involves three key steps:

1. **Recognizing the Emotion:** Taking a moment to notice and acknowledge what we are feeling.
2. **Labeling the Emotion:** Giving it a specific name (e.g., instead of just saying "I feel bad," identifying it as "I feel frustrated," "I feel anxious," or "I feel overwhelmed").
3. **Processing the Emotion:** Once the emotion is named, it becomes easier to manage through techniques like self-soothing, problem-solving, or seeking support.

This process helps integrate emotional experiences into conscious awareness, allowing us to engage with them in a more balanced and productive way. This self-management strategy supports an intentional response to a challenging stimulus rather than an automatic reaction.[16]

Expand Your Emotional Vocabulary

One major obstacle to accurately naming emotions is having a limited emotional vocabulary. Many people rely on a few generic words like "happy," "sad," "angry," or "stressed" to describe their emotions. However, emotions are complex, and having the right words can help us process them more effectively.

Here's an example of how to expand emotional vocabulary beyond basic terms:

Basic Emotion	More Precise Terms
Sad	Disappointed, lonely, heartbroken, grieving, hopeless
Angry	Frustrated, resentful, irritable, bitter, annoyed
Happy	Content, excited, grateful, proud, relieved
Fearful	Anxious, nervous, overwhelmed, vulnerable, hesitant
Stressed	Overwhelmed, pressured, burnt out, restless, tense

A useful exercise is to use an emotion wheel or list of feeling words to refine your ability to pinpoint emotions. The more precise you are, the more control you gain over your reactions.

Resource: Self-Care: Name It to Tame It | Brown Med-Peds (https://brownmedpedsresidency.org/self-care-name-it-to-tame-it/)

Emotional Intelligence: EI versus EQ

EI is also referred to as "EQ" in many spaces of literature and training. In Blue EQ's Leading with Emotional Intelligence Training, the content creators state: "emotional intelligence is the conduit of influence and collaboration." Their motto for the "why" of being more emotionally intelligent is that "EQ is the delivery system for your IQ."[17]

Psychological Safety + Emotional Intelligence

The term "psychological safety" was first coined in 1990 by psychologist William Kahn. Known as "the father of employee engagement," Dr. Kahn's description of this concept was "being able to show and employ oneself without fear of negative consequences of self-image, status, or career." As a blend of trust and respect, this is an environment where people feel safe enough to overcome interpersonal risks. Individuals feel empowered to speak up and share concerns, questions, or ideas.[18]

Elements that have come to be considered essential as part of a psychologically safe climate include:

- Being self-aware of one's impact
- Displaying vulnerability to acknowledge mistakes and repair relationships
- Listening to understand and having judgment-free conversations
- Accepting feedback provided by teammates and leaders
- Putting a stop to blaming others while maintaining accountability
- Acknowledging and validating people's emotions and experiences
- Resolving conflict together instead of allowing it to hold the team back
- Taking a risk, speaking up, and challenging the status quo
- Less cortisol generated out of personal interactions and more oxytocin
- Support for others

Professor Amy C. Edmondson continued the work of Dr. Khan. Now a well-known researcher, author, and speaker on psychological safety, she first explored the impact of psychological safety on work teams in the 1990s. As a first-year doctoral student at Harvard University, she was investigating whether high-performing medical teams made more or fewer mistakes than low-performing teams. In the 1999 publication of her dissertation, she detailed historical and evolving understanding of organizational learning. Examples of learning associated with organizational work teams include seeking feedback, sharing information, asking for help, talking about errors, and experimenting.

Box Out: Psychological Safety as Defined by the Founding Experts

- "Psychological safety is the belief that one will not be punished or humiliated for speaking up with ideas, questions, concerns, or mistakes. It is the shared belief that interpersonal risk-taking is safe."[19]

  –Amy C. Edmondson (The Fearless Organization; https://fearlessorganizationscan.com)
- "Employees feel included, safe to learn, safe to contribute, safe to challenge the status quo without fear of being embarrassed, marginalized, or punished."[20]

  –Timothy R. Clark

Per Dr. Edmondson, when teams lack psychological safety – that is, when they feel unsafe speaking up, asking for help, raising questions, or flagging problems – employees are much more likely to experience workplace stress and, eventually, the emotional and cognitive exhaustion that contributes to burnout. These elements contribute sizably to disengagement and staff turnover.

In both human and veterinary medicine care environments, emotional intelligence and psychological safety have been studied and proven crucial to:

- Foster collaboration, trust-building, and skillful communication, which are essential for highly effective and innovative teams
- Elevate the quality of medicine provided and thus patient safety and outcomes
- Support the holistic health of individual veterinary professionals, teams, and organizations

Psychological Safety: Strategies and Implementation for Veterinary Environments

In all veterinary practice environments, veterinary professionals continuously manage interpersonal risk with fellow caregivers and leaders while concurrently navigating the dynamics of providing quality medical care. The fear of appearing ignorant, incompetent, or disruptive shuts people down. In many instances, this may not even be conscious behavior. Historically, the tendencies of perfectionism in many veterinary care providers support the internal critical and disruptive dialogue associated with imposter syndrome.

In 2024, AAHA supported the Veterinary Visionaries Task Force in the creation of the *Psychological Health and Safety Guidelines*. The 11 Foundational Guidelines were created to support "positive actions leaders of veterinary organizations can take to create environments where they, their employees, clients/patients, and industry partners experience well-being." The guidelines are organized into three core sections: Organizational Culture, Workplace Experience, and Relationships. Each section goes into depth on important tenets that provide clear definitions of terms and principles, examples of interventions and strategies, and existing frameworks that can be utilized to guide additional understanding and implementation.[21]

4 Stages of Psychological Safety

In Dr. Timothy R. Clark's "The 4 Stages of Psychological Safety: Defining the Path to Inclusion and Innovation," he proposes that when leaders cultivate psychological safety, teams and organizations progress through four successive stages. As a result, there is increased confidence and comfort in speaking up and making valuable contributions to work-place discussions. The 4 stages of psychological safety developed by Dr. Clark are:

1. **Stage 1**: Inclusion safety – basic human need of connecting and belonging.
2. **Stage 2**: Learner safety – feel safe to learn, ask questions, and experiment; feel open to give and receive feedback; mistakes can be shared and learned from collectively without fear of judgment, humiliation, or retribution.
3. **Stage 3**: Contributor safety – individuals feel safe to make valuable contributions using their unique skills and gifts.
4. **Stage 4**: Challenger safety – Individual or team feels safe enough to challenge the status quo when an opportunity for change or improvement is identified.

As efforts are made to mitigate fear, install true performance-based accountability, and create a nurturing environment that allows people to be vulnerable as they learn and grow, they will perform beyond your expectations. He proposes that:

> In the workplace, team psychological safety must be a top priority if businesses want to create a successful enterprise. And, more importantly, psychological safety contributes to an inclusive, diverse, and accepting workplace. A workplace where team members feel safe to express themselves.[22]

Practices That Support Emotional Intelligence Building and Psychological Safety

- Pause and name emotions.
- Reflect on emotional triggers.
- Invite feedback and listen to learn.

Strategies

- Veterinary Visionaries Psychological Health & Safety Guidelines – there are so many fantastic ideas for individuals and for teams here!
  - Available online: VV-Psychological-Health-Safety-Guidelines-June-2024.pdf (https://veterinaryvisionaries.org/)

- Use active listening to create a space of safety, trust, and emotional regulation.
- Promote emotionally attuned questions.

Phrases to Use

- "What emotions are showing up for you?"
- "Let's take a breath."
- "I feel overwhelmed at this moment because I ____" (fill in the blank and ask for support!).

Reflections

- What helps me stay grounded emotionally?
- When I feel (name the emotion), I know that I tend to (what do you think/say/do when you experience this emotion), and I know that (supportive strategies) will support me in returning to a state of presence and allow me to respond from my values and higher mind.
- How am I as a leader contributing to a culture of psychological safety to support voice, choice, contribution, and learning for my team and my organization?

Workbook Prompt: Take a self-assessment and a team assessment. Where can you grow?

Resources to support this exercise

- Iterum Ltd.: www.psychsafety.co.uk/measure-psychological-safety/
- The Fearless Organization: Explore Our Tools to Foster Psychological Safety (https://fearlessorganizationscan.com/engage)

Section 4: Resilient Leadership

Complex systems exist widely in nature and in society. Complex systems and network theory have been widely applied to many fields: social networks, information networks, technological networks, and biological networks. Examination of the ability of complex networks to respond to change and network resiliency as a concept has been evolving and rising in significance over the last 50 years. Evolvability and adaptability are the fundamental tenets of resiliency in a system. Let's examine how these concepts may be applied to veterinary leaders, teams, and practices.

Definition

The ability to navigate uncertainty, adapt to adversity, and model emotional agility.

Resiliency refers to the ability of an individual or team to face, adapt, and bounce back from challenges and adversity. There is the potential to foster mental strength, effective stress management skills, and a positive mindset that can help reframe difficult experiences as growth opportunities.

Strong relationships, feeling psychologically safe, and having effective problem-solving skills support individuals and communities in improving overall wellbeing and coping abilities.

4 Types of Resilience

Physical: How our body deals with change and recovers from physical demands, illness, or injury.

Supportive Practices: Getting enough sleep, eating a nutritious/balanced diet, and engaging in regular exercise/movement.

Mental/Psychological: Our capacity to mentally adapt and cope with change, uncertainty and adversity.

Supportive Practice: Utilizing grounding practices to settle the mind and nervous system by calling in the parasympathetic and/or the polyvagal nervous system to bring a sense of calm and invite the problem-solving, perspective-taking prefrontal cortex back online.

Emotional: How we regulate our emotions during times of stress without becoming overwhelmed.

Supportive Practice: "Name it to tame it!" Practicing the noticing and naming of what feelings and emotions are present, what physical sensations are occurring, and what you know to do (tapping into inner wisdom and healthy coping strategies) that will complete the stress cycle and bring you back to your centered, capable self.

Box Out: Wise Words from Emily and Amelia Nagoski from Their Collaborative Book, *Burnout: The Secret to Unlocking the Stress Cycle*

- Just because you have dealt with a stressor does not mean that you have dealt with the stress (physiological impacts of the stressor on you).
- Wellness is not a state of being but a state of action. It is the freedom to move fluidly between the cyclical, oscillating experiences of being a human.

- "Human Giver Syndrome" is the contagious false belief that you have a moral obligation to give every drop of your humanity . . . in support of others, no matter the cost to you.
- Humans are not built to function autonomously; we are built to oscillate from connection to community to autonomy and back again. Connection is as necessary as food and water.[23]

Social/Community Resilience: The ability of groups to recover from difficult situations. Humans are wired for community and social support, particularly during challenging circumstances – connecting with others and working together to solve problems that affect people both individually and collectively.

Supportive practice: Discussing things with others can help an individual or a community gain insight into the challenges they are facing, shift perspective on the circumstances, and possibly come up with new ideas for managing them. This is the power of collective intelligence and resilience.

Signs of Resilience

- **Effective emotional regulation:** In the face of stress, recognizing and feeling both negative and positive emotions without judgment and recognizing that the emotions are temporary and can be managed.
- **Self-compassion:** Demonstrating self-acceptance and kindness toward oneself, especially when circumstances are challenging, when we are in a space of "not-knowing-yet," or when mistakes are made.
- **Feeling in control:** Having a strong internal locus of control supports the feeling that your actions can play a part in determining the outcome of events (releasing "victim mindset").
- **Confidence in problem-solving** and past strengths-building experiences to support effective coping skills.
- **Social support:** Having a solid network of supportive people and knowing when you need to ask for help/support.[24]

Why It Matters

- Resilience is a vital leadership skill as leaders navigate unexpected problems and are better able to handle a crisis.
- Resilient leaders tend to be more socially competent, have better problem-solving skills, and have a better sense of purpose and future.
- Prevents burnout and creates a confident, hopeful team climate.

Box out: Culture Shift

We need a community and culture that meaningfully supports recovery and completion of the stress cycle rather than the current dominant model of rewarding endless work and devaluing rest and recovery.

Protective Factors to Support Individual and Community Health and Resilience

Definition: Mediating processes that buffer the effects of adversity so that health, wellbeing, and performance are sustained in the long term.

Why It Matters

Leaders already have a full plate between self- and team management, and this is without consideration of what may be happening in their personal lives.

> Leaders have a pivotal position in positively affecting the environment and challenges of the workplace as they can influence interventions that impact upon the resilience and wellbeing of others. Leader stress, their affective wellbeing and the styles and the behaviours they adopt are directly associated with employee stress and wellbeing.[25]
>
> – Grey and Jones (2018)

Promoting Protective Factors to Help Leaders and Caregivers Stay Well

1. Buddy/support system (help with managing workloads and emotional support)
2. Reflective supervision
3. Experience = in job wisdom (increase competency and confidence)
4. Wellness practices (individualized; need the time to recalibrate)
5. Training/risk reduction
6. Compassion satisfaction ("goosebump moments," "moments of awe," "you made a difference . . .")
7. Having a supportive, safe, inclusive work environment[26]

Promoting Protective Factors to Support Managers/Leaders in Staying Well

- **Supportive relationships:** Having strong social networks inside and outside of the workplace.
- **Autonomy:** Being empowered to solve problems and take appropriate action in line with practice/organization values and goals.

- **Make a wide variety** of health support and wellness resources available and easy to find for the entire team, including leaders.
  - Employee Assistance Programs/Plans
  - Social workers
  - In-practice wellness champions/ambassadors
  - Partner with an organizational mental health community such as TogetherAll, Lyra or other mental health platforms for professionals
  - Veterinary-created resources/trainings/initiatives (VetLife, THRIVE, AVMA's Wellbeing in the Workplace, Veterinary Hope Foundation, CVMA's abundant wellbeing resources)
- **Supportive leadership strategies** (peer support, professional coaches, professional development opportunities)
- **Life-work balance** (practicing healthy boundaries)
- **Professional learning** and development opportunities

Practices

- Cognitive reframing to shift your perspective, for example, "What's one thing I can take away from this experience?"
- Seeking support. Strong relationships, personal and professional, are the foundation of resiliency. Having trusted friends, colleagues, family members, or even professional support who can provide encouragement and perspective during difficult moments is effective and efficient.
- Focus on what is within control – what are small, achievable steps to focus on when things seem overwhelming?
- Manage stress in the moment (e.g., polyvagal breathing to increase parasympathetic nervous system modulation of the sympathetic system's reactions; progressive muscle relaxation; exercise; expressive writing).
- Practice gratitude daily. This habit shifts your focus to the positives that are also occurring in your life and fosters hope and optimism.

Strategies

- Use stress and emotional regulation tools.
- Normalize failure as learning (growth mindset).
- Encourage recovery breaks (rest comes in many forms and is essential for holistic health as well as productivity, creativity, and knowledge integration).

Phrases to Use

- "Let's focus on what we can control."
- "What growth can we find in this?"
- "What is one step we can take today to feel more empowered?"

Reflection

What helps me bounce back when experiencing stress, overwhelm, or challenging circumstances?

Workbook Prompt

Identify a resilience habit that you feel you can strengthen. What are the ways that you can increase your skills and integrate this resiliency skill into your leadership approach?

Section 5: Collaborative Leadership and Delegation

In many veterinary environments there remains the belief that a "qualified" veterinary professional requires a stoicism, or toughness, to "roll with the punches" – if someone does not feel safe speaking up, that this is also "their" problem, or a lack of fit with the vet med team ethos. A culture of silence is one with significant consequences for individual associates and potentially

Figure 4.1 Resilience habits.

for veterinary patients in their care. When team members do not feel safe, they may adjust their behaviors and responses. This results in a less effective and less communicative team. There is an increase in disengagement and unproductive behavior as well as potential for medical errors and patient risk. Encouraging a culture where it is seen as not only safe but valued to share concerns or ideas is essential to fostering collaboration, improving patient care, and reducing medical errors and possibly deaths. [27]

Psychological safety for the support of the veterinary associate cannot be understated. Without this, every other aspect of the veterinary professional's career is sabotaged. This is important in creating an environment where a colleague is compassionately supported when they are struggling in any capacity. The ability to share when a mistake is made and not fear negative repercussions or shaming is vital to the wellbeing of the individual and in contributing to the growth of knowledge and healthier culture of interpersonal communication for all. This is particularly valuable to new leaders (new to leadership or new to the team). Knowing that the practice culture is one of trust and of respect would support a collaborative approach toward learning as well as putting in safeguards to support autonomy while nurturing associate growth, development, and engagement.

Definition

Collaborative leadership is a powerful approach that emphasizes teamwork, shared responsibility, and the value of each team member's input to achieve common goals.

Shared power, mutual respect, and trust-driven task assignment are all supported in this style of leadership.

Collaborative leadership is a style that involves engaging all team members in the decision-making process. It is characterized by the following principles:

- **Shared Decision-Making:** Leaders actively seek input from team members, valuing diverse perspectives and insights. This approach fosters a sense of ownership and commitment among team members, as they feel their contributions are recognized and valued.
- **Open Communication**: Effective collaborative leaders encourage transparent communication, creating an environment where team members can freely share ideas, concerns, and feedback. This openness enhances trust and strengthens relationships within the team.
- **Collective Problem-Solving**: By leveraging the strengths and expertise of all team members, collaborative leadership enables teams to tackle complex challenges more effectively. This collaborative effort often leads to innovative solutions and improved outcomes.[28]

Box Out: Important Consideration in Building Collaborative Relationships at Work

It is essential to manage expectations by proactively sharing with individuals and teams what impact their voice or vote has on a final outcome when collaborative discussion occurs. Transparency and courageous clarity are essential to avoid over-promising and underdelivering. Without this, leaders can lose trust and progress in achieving collaborative relationships and a culture of safety and honesty.

Why It Matters: The Benefits of Collaborative Leadership

- **Enhanced Team Dynamics:** Collaborative leadership fosters a positive team culture where members feel valued and respected. This inclusivity leads to higher morale and job satisfaction.
- **Increased Innovation:** When team members contribute their unique perspectives, it can spark creativity and innovation, resulting in better problem-solving and more effective strategies.
- **Improved Performance:** Teams that operate under collaborative leadership often achieve higher performance levels, as shared goals and collective accountability drive motivation and commitment.
- **Increased Adaptability:** Collaborative leadership allows practices and organizations to be more agile and responsive to new challenges, as teams can quickly mobilize their collective resources, skills, and insights.
- **Delegation** empowers growth and relieves leadership burden. This allows an excellent opportunity for a teammate to test out their leadership abilities, which builds confidence, supports engagement, and fortifies the relationship.
- **Cultivates** a resilient and self-sufficient team.

Practices That Support Collaborative Leadership and Delegation

- **Match** people to tasks.
- **Set** clear expectations.
- **Encourage Participation:** Create opportunities for all team members to contribute to discussions and decision-making processes. This can be achieved through brainstorming sessions, workshops, or regular team meetings.
- **Build Trust:** Foster a culture of trust by being transparent, reliable, and supportive. Trust is essential for open communication and collaboration.
- **Provide Training:** Equip team members with the skills needed for effective collaboration, such as communication, conflict resolution, and teamwork strategies.

- **Celebrate Successes:** Acknowledge and celebrate team achievements to reinforce the value of collaboration and motivate continued teamwork.

Key Takeaway
By embracing collaborative leadership, organizations can create a more engaged, innovative, and high-performing workforce that is well-equipped to meet the challenges of today's dynamic environment.

Collaborative Leadership: Embrace the Art of Delegation

> "Trust is the glue of life. It's the most essential ingredient in effective communication. It's the foundational principle that holds all relationships." [29]
>
> – Stephen R. Covey

Delegating Effectively Requires Building and Keeping Trust

3 Dimensions of Trust

1. **Trust of Character**
   Trust of Character establishes the foundation for teamwork by ensuring members believe each other will act as promised. This trust forms when team members consistently keep their commitments, fostering mutual reliability and effective collaboration.
2. **Trust of Communication**
   Trust of Communication encourages open dialogue and safe direct interactions among team members. It involves not just sharing information but also addressing concerns and giving constructive feedback for learning and growth. With transparent communication, teams foster honesty, admit mistakes, and know their standing, creating a collaborative environment vital for success.
3. **Trust of Capability/Competence**
   Trust of capability enables team members to contribute their expertise, collaborate in decision making, and share skills. This trust fosters innovation, resilience, and engagement.[30]

The Art of Delegation: Step-by-Step

1. **Understand your *preferences* for delegation.**
   Leaders who delegate effectively have prioritized their workload and know why they're handing off some tasks and holding on to others. They also know themselves and what they want. For example, do you want constant updates or just an update now and then? Will you let the person, or team, assigned the task determine the process, or must you have it done your way? When you're clear

about your preferences, you can better communicate those needs and expectations to other people.

2. **Understand your *people* and match tasks to strengths**
   Effective delegation involves assigning people tasks, responsibilities, and duties that match their knowledge, skills, abilities, and interests. It involves giving assignments that will help them learn to lead and grow. By truly understanding the people you manage, you can effectively identify the individuals to whom you should (or should not) delegate specific tasks.
3. **Understand the task and its *purpose*.**
   The task is what must be accomplished. The purpose is the reason that the task is being done – it gives meaning to the task. When a leader effectively matches the purpose of a task with a team's or individual's beliefs and goals, it becomes an opportunity for development. This is the idea behind purpose-driven leadership, which helps increase employees' commitment and willingness to go beyond the original assignment if necessary to accomplish the purpose.
4. **Provide clear instructions and expectations**
   When delegating, be sure to explain your expectations and share timelines/deadlines and the desired outcome. If there are additional criteria, such as a specific protocol for the individual to be aware of, this should be communicated up front. Being clear in your instructions minimizes mistakes and frustration and sets your team members up for success.
   Note: Ask about, and allocate, necessary resources and support needed to complete the tasks (time, information, additional staff members, open communication with you as needed).
5. **Trust your team and "Let go!"**
   This can be one of the biggest hurdles in delegation, and it is such an important space for building interpersonal relationships. It is natural to want to stay involved and in control. Micromanaging only defeats the purpose of delegation in being an energy drain on the manager/leader and can stifle growth in the team member. Be available for support but provide them with the autonomy to do the work and build their skills and confidence as a result. Remind yourself and others around you to have grace while others learn by doing.
6. **Share the *process* for assessing and rewarding accomplishment of the task.**
   What does successful accomplishment look like? How will you be assessing the task itself and how it's accomplished? Will you require

people or teams to report to you after every step of the project, or will you just ask for a weekly rundown? Will you develop a series of benchmarks you want them to meet with regard to the project? And when it's over, or at key points along the way, can you find a way to recognize and reward the people and teams who accomplish what you asked them to do? Feedback and recognition are essential for growth and for building confidence in your team members.[31]

Box Out: Failing Forward

In order for delegation to be effective and impactful, consider giving room for "failing forward." Failing forward is the reframing of failure. Failure is not something to fear or to try to avoid but is an intrinsic part of learning, exploration, and building confidence. If the desire to feel safe and to belong promotes repeating "safe recipes" that were good enough in the past, innovation and growth are stifled. If individuals are anticipating humiliation or punitive actions if a mistake is admitted to or clarity is needed on an assigned task, psychological safety is not present. The advancement of an individual's professional development is resultingly significantly limited.

In medical caregiving environments, open communication and collaboration are essential. Enabling healthy, supported discussion of reasons for mistakes or for errors allows for identification of the risk factors that need to be addressed and for new, improved protocols and policies to be developed.

Purposeful Delegation

A strategy to demonstrate real affirmation, creating an environment that regular provides people the evidence of their significance and meaning as part of the team and organization:

- **Significance:** Show how the task impacts other people.
- **Necessity:** Describe what the task will make possible.
- **Strengths:** Remind people which strengths they can use to do the task.[32]

Key Takeaways

- Effective delegation can play an important role in the functioning of a veterinary practice or organization.
- By assigning tasks according to team members' strengths and providing clear instructions, it is possible to enhance workflow and encourage collaborative work.

- Delegation may also contribute to a more engaged team environment, which can support consistent care for patients and clients.

Phrases to Use

- "This plays to your strengths."
- "I need you to work on (task) and complete it by (deadline)."
- "This task is important because (reason). Can you take responsibility for it?"
- "This is what success looks like for this task"
- "You can reach out to me if anything is not clear"
- "Do you need any additional resources to complete this?"
- Let's discuss how we can make this even better."

Reflections

- What makes delegation feel safe for me?
- What are my limiting beliefs about delegation?
- What are the potential personal barriers I can prepare for when delegating a task?
- What strengths of key performers on my team can make for successful delegation of a task that is on my plate currently?

Workbook Prompts

- Identify a task to delegate.
- What are some of the specific phrases that you might use when delegating this task?
- How will the delegation of this task benefit you? The team? The practice/organization?

Collective Intelligence and Resilience: "We" are Smarter and Stronger Than "Me"

> "Human knowledge is never contained in one person. It grows from the relationships we create between each other and the world, and still, it is never complete."
>
> Paul Kalanithi, *When Breath Becomes Air*[33]

Collective intelligence is the body of knowledge that grows out of a group. When groups of people work together, they create intelligence that cannot exist on an individual level. Components of collective intelligence include:

- Making decisions as a group
- Forming a consensus
- Getting ideas from different sources
- Motivating people through competition

Thomas Malone, the Patrick J. McGovern Professor of Management at MIT and the founding director at the MIT Center for Collective Intelligence, coined the term. In his book *The Future of Work*, Malone theorizes that businesses of the future will look very different from today's organizations because of collective information gathering. Artificial intelligence involving computers and other types of automation have produced collective intelligence as different types of groups connect to produce a body of knowledge.[34]

Interprofessional Competencies and Effectiveness

Interprofessional competencies are essential for *effective collaboration* in health care and other fields.

Key competencies include:

- **Values and Ethics**: Understanding and respecting the values of different professions.
- **Roles and Responsibilities**: Clarifying the roles of each team member in collaborative practice.
- **Interprofessional Communication**: Effectively sharing information and ideas across disciplines.
- **Teamwork and Team-Based Care**: Working collaboratively to achieve common goals.

Additional competencies may include:

- Role clarification
- Team functioning
- Patient/client/family/community-centered care
- Interprofessional conflict resolution
- Collaborative leadership.[35]

Collective Resilience

Collective resilience is a new and emerging concept that originated in the moral and ethical context of emergencies and disasters in seeking a new perspective on the aspects of personal and individual resilience and recovery in terms of "shared identification, solidarity, and cohesion."[36]

This work – the work of becoming more conscious in our relationships, of learning to recognize our defensive patterns before they hijack our nervous systems – feels both deeply personal and essentially collective. As a personal friend and mindset coach said to me recently: "We cannot do this work in isolation. We need witnesses, mirrors, fellow travelers who can hold us accountable to our highest intentions."[37]

The path toward relational healing and collective resilience in the face of difficulty requires:

- Courage to examine our reflexes, beliefs, biases, and assumptions
- Compassion for our imperfections (aka, we are human!)
- Community to support our growth

It asks us to consider where and how we individually and collectively own and act on those experiences in our veterinary environments that may cause distress.

As an example, we can individually and collectively build moral resilience and ethical competency in our caregiving environments where moral distress and ethical conflicts occur regularly. It has been said that "moral resilience is the antidote for the powerlessness felt from moral distress." Here are some considerations on how we can shift toward possibility and choice:

- Define/refine your personal moral compass.
- Define a personal code of ethics.
- Work on self-awareness and self-regulation as part of strengthening your EQ.
- Lean into connectedness and support.
  - Engage with compassionate community.
  - Find meaning in the midst of distress or despair.
  - Identify and build relationships with morally resilient mentors.

Key Takeaway

Resilience is both a personal and a community endeavor. Reach out and leverage your personal and professional connections as a safety net when you struggle to address complex ethical situations. This helps to support your overall mental health and ability to emotionally regulate when morally distressed and perhaps experiencing a sense of moral disappointment. In this you and your community are better able to release, to the extent that is possible, the moral residue and return to a healthier state of homeostasis.[38]

Section 6: Systems Change and Culture Evolution

Embracing the Multidisciplinary Approach in Veterinary Medicine

There is literally a world of wisdom that is available to veterinary leaders and culture change agents. Borrowing and translating interprofessional wisdom is smart, efficient, and effective. Effective strategies have been created through years of research and application in many other professions and can be thoughtfully integrated into the needed solutions to address the current global challenges facing the veterinary industry.

An important shift for veterinary teams to embrace is examining issues as interconnected in a systems approach. In particular, there has been a progressive conversation from both human medicine and veterinary professional environments supported by research to move from reactive mental-health maintenance to a system-wide, proactive wellbeing ecosystem. These same tenets apply to the "wellness" of the entire ecosystem, whether at the practice level, organization, or for the entire veterinary industry.

Definitions

- **Culture** is a set of shared behaviors and beliefs shaped by leadership and reinforced over time. Sociologists define the non-material aspects of culture as the values, language, communication, and practices that people share and uphold in common and that can be used to define them as a collective. Knowledge, expectations, rules, and morals contribute to culture and inform how a community behaves.
- **A Multidisciplinary approach** involves collaboration among professionals from different disciplines to address complex problems and improve outcomes, particularly in fields like health care, education, and research.
- **Multidisciplinary teams (MDT)** – considering the individual talents and strengths of team members within a practice or organization and combining them in such a way that exponentially increases the abilities and capabilities of the individual. This increases efficiency and innovation, decreases the "us" and "them" mentality, and embraces the unique contributions of individuals supporting engagement, belonging, and purpose-driven, collaborative decision-making.

Why It Matters

- An individual's responsibility, resiliency, and actions alone is not enough – meaningful change requires systemic transformation at team and organizational levels.

- Implementation comes with challenges: need for leadership buy-in, proper training, institutional commitment, and longitudinal evaluation to ensure sustainability as well as appropriate innovation/adaptation to evolving circumstances of the business, industry. and world.
- Systemic shifts that embrace effective multidisciplinary knowledge and strategy as well as veterinary-centric approaches are essential for sustained veterinary wellness and for a healthy veterinary profession going into the future.

From Social Science: I Frame vs S Frame

Note: The i-frame refers to science and policy focused on changing individuals. The s-frame concentrates on everything around the individual (the larger ecosystem).

The behavioral and brain sciences primarily focus on what we call the i-frame: on *individuals* and their thoughts and behaviors. Public policy, by contrast, typically focuses on the s-frame: the *system* of rules, norms, and institutions usually studied by economists, sociologists, legal scholars, and political scientists.

In their paper, Chater and Lowenstein elaborate on the differences between interventions that focus on individual behavior (i-frame), as opposed to the systems in which health behavior occurs (s-frame). They point out that behavioral scientists frequently rely on individual-level interventions rather than systemic change to improve population health.[39]

From many professions globally, there is mounting evidence that individual-level interventions are less effective than structural interventions in supporting occupational wellness for workers and the sustainable health of an organization/business.

From Positive Psychology Theory: Systems-Informed Positive Psychology (SIPP)

Systems-Informed Positive Psychology (SIPP) incorporates principles and concepts from the system sciences into positive psychology theory, practices, discourse, and methodologies to optimize human social systems and the individuals within them.[40]

SIPP returns to the wisdom of Aristotle, beginning "to construct a science (both basic and applied) of the good life that has an eye to the subtleties and complexities of change in human systems."[41] Instead of solely fixing problems reactively (e.g., mental health support for teams via resilience workshops or employee assistance), there is growing research that advocates for this **systems-informed positive psychology** framework – also called **positive**

education – to be applied at **individual, team,** and **organizational** levels in veterinary training and practice environments.

Specifically, when addressing wellness in the workplace, the PERMA + 4 model has been demonstrated to be effective as a science-backed approach to improve work-related wellbeing including engagement, performance, and stress levels. The PERMA model was raised as a topic in Chapter 2. PERMA + 4 extends this to work by including physical health, mindsets, work environment, and economic security. Utilizing available assessment tools can highlight areas for potential growth and development. Taking consistent, small steps to implement some of the tools associated with this model can contribute to sustainable impact on a culture of health and wellness for a team and organization.[42]

Practices to Bring a Multidisciplinary Approach to Healthy Culture Creation

- Define organization and team norm and identify values.
- Be intentional and strategic in your practice and/or organization's culture evolution by using tools like "Start, Stop, Continue"[1] when examining policies and protocols.
- Model the culture you want by conducting regular culture assessments.

There are various models an organization may use to evaluate their internal and external cultures. Some organizations choose to perform culture walks, interviews or surveys. One of the most popular cultural assessment tools used throughout various industries is the **Organizational Culture Assessment Instrument (OCAI)** developed by researchers Kim Cameron and Robert Quinn at the University of Michigan. The OCAI is a useful survey model that helps organizations develop an awareness of their culture, understand its inner workings and identify what cultural components they'd like to change.[43]

Strategies to Support a Culture of Wellness and Collaboration

- Co-create team agreements.
  - Examining current and future strategies and policies – use the lens and language provided from values/vision/mission and team agreements.
  - Accountability is not a bad word! It is essential for a culture built on trust and shared responsibility.

  - Consider and discuss how leadership and the staff will communicate and behave to demonstrate the team agreements in the workplace.
- Use PERMA + 4 audits, such as *the Positive Functioning at Work Assessment*, before and while implementing the associated evidence-based practices, such as:
- Break down large goals into smaller milestones to gauge and celebrate progress.
- Take walking breaks and meetings throughout the workday.
- How can you modify your workspace to support your focus, mental health, and physical health (e.g. ergonomics)?
- Promote financial literacy for yourself and for your team to increase feelings of economic preparation and security.

Phrases to Use

- "How does this reflect our values?"
- "What is our shared mission and purpose at this practice/organization?"
- "What are we currently doing to make sure that these values are known and upheld by our community, that is, what is in place to incorporate a growing team into the team norms?"

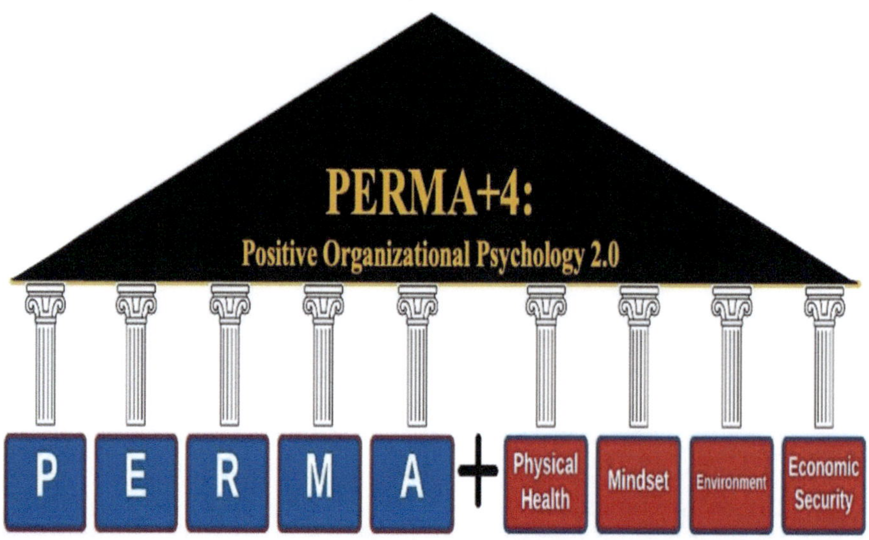

Figure 4.2 The PERMA + 4 Framework: A New Approach to Well-Being at Work
Psychology Today **article The PERMA+4 Framework: A New Approach to Well-Being at Work |** *Psychology Today*.
Source: Donaldson, Van Zyl, and Donaldson (2022).

Reflections

- What value is missing from our culture?
- What do we want to be known for as a team? As a business/organization?
- Do our clients know and experience our values? How are we measuring this?
- How do I/we view setbacks and challenges? How can we practice a growth mindset more regularly?

Workbook Prompt: Sketch your ideal team culture.

Box Out: Multitiered Intervention Strategies

1. **Individual (Me):**
   - Tools like *strengths assessments, mindfulness, wellbeing literacy,* and *character-strength exercises* to build personal resilience and flourishing.
2. **Team (We):**
   - Foster *high-quality connections, peer support, mentoring,* and *experiential learning* to enhance *engagement, belonging,* and *support networks.*
3. **Organizational (Us):**
   - Cultivate a *positive culture, effective leadership, improved physical workspaces, fair compensation, schedule flexibility,* and *alignment with core values* to support wellbeing at scale.[44]

Frameworks for Building a Culture of Wellbeing

There are several global communities inside and outside of veterinary medicine that have created frameworks to provide support for health care workers' wellbeing.

In the United States, The National Academy of Medicine has created a National Plan for Health Workforce Wellbeing as a framework with seven priority areas. (https://nam.edu/publications/national-plan-for-health-workforce-well-being/)

In Australia, Safe Work Australia developed the model work health and safety (WHS) laws to be implemented across Australia. Each state and territory is responsible for implementing and regulating these laws. Additionally, the National Safety Council of Australia (NSCA) produced the AR Interactive – Veterinary Industry Guide to Workplace Safety 2023/24 in conjunction with the Australian Veterinary Association. (https://www.provisual.com.au/veterinary-industry-guide-to-workplace-safety/)

In the United Kingdom, the British Veterinary Association (BVA) has created "Great Workplaces," an accreditation scheme designed to recognize and reward positive workplace culture across both clinical and non-clinical veterinary workplaces. In addition, the BVA has included policy position details, resources, and webinars on "good workplaces" for veterinary professionals.

(https://www.bva.co.uk/take-action/good-veterinary-workplaces/good-workplaces-health-and-wellbeing/).

On their website they list some of the elements that make for "good workplaces":

- Develop, establish, and track good physical and mental wellbeing and are supportive of the needs of all team members.
- Ensure personal safety always takes precedence, including over professional responsibility.
- Have a culture of respecting personal health rather than one of "pushing through" illness or injury.
- Proactively manage workplace stress.
- Foster and support a culture of psychological safety, in which all team members' mental health and wellbeing are supported.
- Recognise the potential risks posed by lone working and domiciliary visits and take steps to mitigate those risks.[45]

Additionally, Deloitte, a British multinational professional services network, has a wide variety of resources and strategies available that would apply to a myriad of professions worldwide. Helpfully, here are some suggestions from the Deloitte community for leaders to think through as to what changes they can make in 5 environments across the 3 levels (organization, team, individual).

Cultural

Building Wellbeing into Social Behaviors and Norms

Organizations: Model wellbeing behaviors such as taking micro-breaks or only making certain meetings video-focused

Individuals: Be proactive and vocal about wellbeing needs.

Relational

Fostering Wellbeing in Relationships Among Colleagues

Organizations: Form teams based on worker preferences, working styles, and personal needs in addition to skill sets.

Individuals: Check in frequently, proactively, and consistently with colleagues on their wellbeing needs and preferences. (A close friend and skillful veterinary leader shared she calls these "accountabilibuddies!")

Operational

Including Wellbeing in Management Policies, Processes, and Programs

Organization: Embed wellbeing criteria into scheduling, performance management processes, leadership evaluations, and rewards and recognition programs.

Teams: Enable team agency and choice by allowing teams to adopt the wellbeing practices best suited to them.

Physical

Designing the Physical Workspace to Facilitate Wellbeing

Organizations: Design work environments to support workers' physical, mental, and emotional health needs.

Teams: Leverage physical workspaces that promote team collaboration and performance

Virtual

Designing New Technologies and Virtual Workplaces for Wellbeing

Organizations: Use new technologies, like virtual reality, to train team members to navigate stressful situations, for example, interacting with a frustrated client. Create opportunities for remote work roles to support individual workers' health needs and life responsibilities.

Individuals: Leverage wearable technologies and apps to help master distractions, increase mindfulness, and reduce anxiety.[46]

> **Box Out**
>
> "When an organization is able to successfully design wellbeing into work, wellbeing becomes indistinguishable from work itself and embedded across all organizational levels and environments so that it drives and sustains not only human performance but also human potential."[47]
>
> – Fisher and Phillips (2021)

Additional strategies to activate trust-building and support organizational wellness:

- Practice delegation of tasks and projects (great opportunity to promote professional development, build trust, and foster greater team engagement).
- Create a support system (inside and outside of the workplace).
- Design protocols and SOPs to provide clarity and manage expectations.
- Foster an environment of psychological health and safety for all employees.
- Empower the team to solve problems for themselves (*Note: A foundational environment of psychological safety and trust is essential for this to successfully occur*).
- Effective time management (e.g., optimize technology, plan meal breaks, "plan your work and work the plan").
- Practice individualized, impactful self-care (model the behavior for others), and develop healthy coping strategies for work-related stress.
- Cultivate a culture of permission for compassion for the entire team – develop a system of micro-breaks for effective, evidence-based cognitive, physical, and emotional resets during the work shifts.
- Encourage use of PTO to recalibrate and decrease the likelihood of occupational distress and burnout.

Section 7: Conscious Change and Leadership

Drs. Jean Kantambu Latting and V. Jean Ramsey have been researching, teaching, and writing on how to realistically effect personal and organizational positive change for decades. They developed a set of research- and theory-based guidelines for powerful self-reflection, bridging cultural divides, and spearheading change that led to the authors' first book, *Reframing Change: How to Deal with Workplace Dynamics, Influence Others, and Bring People Together to Initiate Positive Change* in 2009.

The evidence-based framework that they created supports the practical benefits of behavioral science research and implementing skill-building strategies at individual, team, and organizational levels: manage emotions better, improve relationships, and resolve conflicts more effectively.[48]

Over the next decade, the authors continued teaching and training these concepts with the Leading Consciously consulting and training organization.

Their team continued to seek empirical research to fine-tune the skills and principles as well as to improve their ability to describe and illustrate them with contextual examples and real-life stories. Their most recent book, *Conscious Change: How to Navigate Differences and Foster Inclusion in Everyday Relationships* (2024), reviews the fundamental concepts and tenets of the conscious change principles and brings them to life through 19 stories shared by individuals who applied the skills in a wide variety of roles, jobs, and employment situations.

"Conscious change," as defined by the authors, focuses less on getting others to change and shifts toward the power of self-change and the resulting more intentional, values-based actions one takes that impact others. There are 6 principles, and each principle has its own set of skills, for a total of 36 skills in all.

Note: I recognize that the language used in the provided list of the 36 skills below may seem unusual when taken out of context of the larger book that explains and applies them. Consider this an invitation to get curious and learn more about these concepts by reading/listening to the book!

Test Negative Assumptions

1. Move from the answer into the question.
2. Look for multiple points of view.
3. Consciously test your negative assumptions.
4. Check to see if you are making cultural assumptions.

Clear Emotions

5. Identify with your values, not your emotions.
6. Avoid emotional suppression.
7. Clear your negative emotions.
8. Build your positive emotions.

Build Effective Relationships

9. Engage in powerful listening.
10. Develop skills in inquiry and openness.
11. Learn how to give, receive, and seek feedback.
12. Distinguish between intent and impact.
13. Apologize effectively.

Bridge Differences

14. Address systemic biases.
15. Learn to recognize dominant/nondominant dynamics.

16. Check for stereotyping tendencies, unconscious bias, and lack of awareness in your behavior, especially as a dominant-group member.
17. Sustain chronic unease toward exclusionary behaviors.
18. As a dominant-group member, provide support to nondominant group members.
19. As a nondominant, resist any tendency toward internalized oppression or viewing dominants as beyond your ability to influence.
20. As a nondominant, recognize dominants' potential unawareness about the impact of their behavior.
21. Call others in rather than calling them out.

Conscious Use of Self

22. Accept responsibility for your own contributions.
23. Maintain integrity.
24. Seek to understand others' perspectives.
25. Focus on others' strengths.
26. Adopt a growth mindset.
27. Recognize your power and use it responsibly.
28. Build resilience through self-affirmation.

Initiate Change

29. Commit to personal change.
30. Emphasize changing systems, not just individuals.
31. Surface undiscussables.
32. Gain support one person (or small group) at a time.
33. Set direction, not fixed outcomes.
34. Learn from resistance.
35. Cultivate radical patience through the time lag of change.
36. Acknowledge small wins.[49]

Key Takeaway

Conscious Change bridges personal insight with organizational impact – providing a toolkit for change grounded in self-awareness and collaboration.

Chapter Summary: Leadership and Culture as Cornerstones of Wellbeing

Veterinary wellbeing is shaped by leadership behaviors and cultural norms. This chapter explored how intentional leadership – rooted in compassion, resilience, and systems-thinking – can drive long-term wellness and satisfaction in veterinary environments.

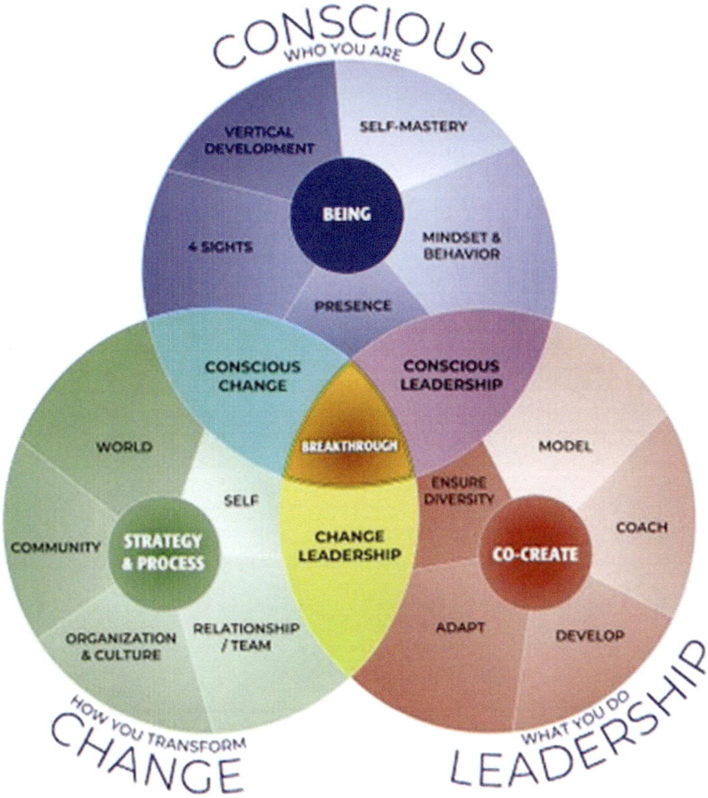

Figure 4.3 Conscious Change Leadership.
Source: https://beingfirst.com/approach/conscious-change-leadership (2020 Being First, Inc.).

Key Takeaways

- Compassion fosters emotional resilience.
- Positive leadership builds safety and clarity.
- EI and psychological safety enable growth.
- Delegation and systems-thinking distribute wellbeing.

Bridge to Chapter 5: From Local Leadership to Global Awareness

Knowing what you know, you have the power to make changes within your practices. You can advocate for yourself and others. We talked in this chapter a lot about the power of systemic change, and this is how it happens. Collectively, we are coming together and advocating for a healthier profession.

Let's discover what the world is learning – and how we can lead change together in the next chapter.

Figure 4.4 Our Wellness Journey now takes us as a more informed collective towards positively impacting the future of veterinary medicine and animal welfare globally.

Note

1 The Start, Stop, Continue method is a simple feedback and reflection framework that helps individuals and teams identify what to start doing, what to stop doing, and what to continue doing to improve performance and outcomes.

References

1. Bennis, W, and Nanus, B. (1997). *Leaders. The Strategies For Taking Charge.* New York: Harper Business.
2. Teambuilding.com. (2025). *Key Leadership Skills & Competences Examples.* https://teambuilding.com/blog/leadership-skills.
3. Gartner. (2025). *Gartner Survey Finds Leader and Manager Development Tops HR Leaders' List of 2025 Priorities for Third Consecutive Year.* https://www.gartner.com/en/newsroom/press-releases/2024-10-15-gartner-survey-finds-leader-and-manager-development-tops-hrleaders-list.
4. Goetz, JL, Keltner, D, and Simon-Thomas, E. (2010). Compassion: An evolutionary analysis and empirical review. *Psychological Bulletin*, 136(3), 351–374. https://doi.org/10.1037/a0018807.
5. Klimecki, OM, and Singer, T. (2012). Empathic distress fatigue rather than compassion fatigue? Integrating findings from empathy research in psychology and social neuroscience. In Oakley, B, Knafo, A, Madhavan, G, and Wilson, DS (Eds.). *Pathological Altruism*. New York: Oxford University Press.
6. Singer, T, and Klimecki, OM. (2014). Empathy and compassion. *Current Biology*, 24(18), 875–878.
7. Cherry, K. (2024). *Maslow's Hierarchy of Needs.* https://www.verywellmind.com/what-is-maslows-hierarchy-of-needs-4136760.
8. Monzani, L, and van Dick, R. (2023). Positive leadership in organizations. In Monzani, L, and van Dick, R. (Eds.). *Oxford Research Encyclopedia of Psychology*. Oxford: Oxford University Press.
9. Wong, CA, and Cummings, GG. (2009). The influence of authentic leadership behaviors on trust and work outcomes of health care staff. Journal of Leadership Studies, 3, 6–23. https://doi.org/10.1002/jls.20104.
10. Tuhin, M. (2025). *The Science of Gratitude. How It Changes Your Mind.* https://www.sciencenewstoday.org/the-science-of-gratitude-how-it-changes-your-mind.
11. Mental Health Commision of Canada. (2018). *National Standard for Psychological Health and Safety in the Workplace (the Standard).* https://mentalhealthcommission.ca/resource/national-standard-for-psychological-health-and-safety-in-the-workplace-the-standard/.
12. AAHA. (2024). *Veterinary Visionaries Psychological Health and Safety Guidelines.* https://www.veterinaryvisionaries.org.
13. Goleman, D. (1995). *Emotional Intelligence: Why It Can Matter More Than IQ*. New York: Bantam.
14. Huberman Lab Clips on YouTube. (2024). *How to Give and Get Constructive Feedback with Dr. Adam Grant and Dr. Andrew Huberman.* https://www.bing.

com/videos/riverview/relatedvideo?q=How+to+Give+and+Get+Constructive+Feedback&&mid=401A9546C677EC052E83401A9546C677EC052E83&FORM=VCGVRP.

15. Bradberry, T. (2025). *The New Emotional Intelligence*. Ottawa: Bruyere Publishing.
16. Mindfulness.com. (2025). *Name It to Tame It: Label Your Emotions to Overcome Negative Thoughts*. https://mindfulness.com/mindful-living/name-it-to-tame-it.
17. Blue EQ's Emotional Intelligence and Psychological Training for Leaders. https://blueeq.com.
18. Kahn, WA. (1990). An exercise of authority. *Organizational Behavior Teaching Review*, 14(2), 28–42.
19. Edmondson, AC. (1999). Psychological safety and learning behavior in work teams. *Administrative Science Quarterly*, 44(2), 350–383.
20. Clark, TR. (2020). *The Four Stages of Psychological Safety: Defining the Path to Inclusion and Innovation*. Oakland: Berrett-Koehler Publishers.
21. AAHA. (2024). *Veterinary Visionaries Psychological Health and Safety Guidelines*. https://www.veterinaryvisionaries.org.
22. Clark, TR. (2020). *The Four Stages of Psychological Safety: Defining the Path to Inclusion and Innovation*. Oakland: Berrett-Koehler Publishers.
23. Nagoski, E, and Nagoski, A. (2019). *Burnout: The Secret to Unlocking the Stress Cycle*. New York: Ballantine.
24. Verywellmind.com. (2025). *How Resilience Helps You Cope With Life's Challenges*. https://www.verywellmind.com/what-is-resiliencee-2795059
25. Grey, D, and Jones, K. (2018). The resilience and wellbeing of public sector leaders. International Journal of Public Sector Leadership, 14(3), 138–154. https://doi.org/10.1108/IJPL-09-2017-0.
26. American Mental Wellness Association. *Risk and Protective Factors*. Accessed online June 10, 2025. https://www.americanmentalwellness.org/prevention/risk-and-protective-factors/.
27. Vaisman, J. (2023). *Lead to Thrive: The Science of Crafting a Positive Veterinary Culture*. Lakewood: American Animal Hospital Association.
28. Kouzes, JM, and Posner, BZ. (2023). *The Leadership Challenge Workbook*, 4th ed. Hoboken: Wiley.
29. Covey, S. (2020). *The 7 Habits of Highly Effective People*, 30th Anniversary ed. New York: Simon & Schuster.
30. Center for Creative Leadership. (2024). *How to Build Trust in the Workplace and on Your Team with Delegation*. https://www.ccl.org/articles/leading-effectively-articles/build-trust-in-the-workplace/.
31. Turregano, C. (2013). *Delegating Effectively: A Leader's Guide to Getting Things Done*. Greensboro: Center for Creative Leadership Press.

32. Mercurio, Z. (2022). *How to Create Mattering at Work*. https://www.zachmercurio.com/2022/11/mattering-at-work/.
33. Kalenithi, P. (2016). *When Breath Becomes Air*. New York: Random House.
34. Malone, TW. (2004). *The Future of Work: How the New Order of Business Will Shape Your Organization, Your Management Style, and Your Life*. Boston: Harvard Business Review Press.
35. Interprofessional Educational Collaborative. (2023). *Core Competencies (Version 3)*. https://www.ipecollaborative.org/ipec-core-competencies/.
36. Delgado, J, Siow, S, de Groot, J, McLane, B, and Hedlin, M. (2021, March 24). Towards collective moral resilience: The potential of communities of practice during the COVID-19 pandemic and beyond. *Journal of Medical Ethics*. https://doi.org/10.1136/medethics-2020-106764.
37. Personal communication, June 17, 2025.
38. Liu, S, Yu, B, Xu, C, Zhao, M, and Guo, J. (2022, November 14). Characteristics of collective resilience and its influencing factors from the perspective of psychological emotion: A case study of COVID-19 in China. *International Journal of Environmental Research and Public Health*, 19(22), 14958. https://doi.org/10.3390/ijerph192214958. PMID: 36429706; PMCID: PMC9690399.
39. Chater, N, and Loewenstein, G. (2023). The i-frame and s-frame: how focusing on the individual-level solutions has led the behavioral public policy astray. *Behavioral and Brain Sciences*, 46, E147. https://doi.org/10.1017/S0140525X22002023.
40. Kern, M., Williams, P, Spong, C, Colla, R, Sharma, K, Downie, A., et al. (2019). Systems informed positive psychology. *The Journal of Positive Psychology*, 15(6), 705–715. https://doi.org/10.1080/17439760.2019.1639799.
41. Hogan, MJ. (2008). Modest systems psychology: A neutral complement to positive psychological thinking. *Systems Research*, 25, 717–732.
42. Donaldson, SI, Van Zyl, LE, and Donaldson, SI. (2021). PERMA+4: A framework for work-related wellbeing, performance and positive organisational psychology 2.0. *Frontiers in Psychology*, 12(817244).
43. Cameron, K, and Quinn, R. *Organizational Culture Assessment Instrument. Available through the Official OCAI Website*. https://www.ocai-official.com.
44. Corrigan, V, Newman, RL, Richmond, P, Strand, EB, and Vaisman, JM. (2025). The future of flourishing in veterinary medicine: A systems-informed positive psychology approach in veterinary education. Frontiers in Veterinary Science, 11, 1484412. https://doi.org/10.3389/fvets.2024.1484412.
45. British Veterinary Association. (2024). *Great Workplaces Accreditation*. https://www.bva.co.uk/take-action/good-veterinary-workplaces/good-workplaces-health-and-wellbeing/.

46. Deloitte. (2021). 2021 *Deloitte Global Human Capital Trends – the Social Enterprise in a World Disrupted: Leading the Shift from Survive to Thrive.* https://www.deloitte.com/us/en/insights/topics/talent/human-capital-trends/.
47. Fisher, J, and Phillips, A. (2021). *Work Better Together: How to Cultivate Strong Relationships to Maximize Well-Being and Boost Bottom Lines.* New York: McGraw Hill.
48. Latting, JK, and Ramsey, VJ. (2009). *Reframing Change: How to Deal with Workplace Dynamics, Influence Others, and Bring People Together to Initiate Positive Change.* Westport: Praeger.
49. Latting, JK, and Ramsey, VJ. (2024). *Conscious Change: How to Navigate Differences and Foster Inclusion in Everyday Relationships.* Scottsdale: She Writes Press.

5

Global Perspectives on Burnout and Occupational Wellness in Veterinary Medicine

Burnout and occupational distress are not new challenges in caregiving professions, but their rising prevalence – and the devastating toll on individuals and systems – have captured global attention. In veterinary medicine, these issues intersect with longstanding cultural norms, economic pressures, and structural inefficiencies. This chapter explores what we can learn from global trends in workplace wellness, how those lessons apply to veterinary practice, and what current research is revealing about the state of our profession.

We'll examine thought leadership, survey data, and organizational case studies, pairing insights with reflective prompts and actionable practices to help veterinary leaders and teams evolve toward healthier, more sustainable ways of working in veterinary care environments.

Section 1: Understanding Burnout – Global Definitions and Nuances

Progress begins with cultivating awareness and shared terminology to identify and facilitate dialogue about challenges at the individual, team, and organizational levels. Veterinary professionals are progressively acquiring the language needed to articulate the emotional impact of caregiving. Given the current high turnover among medical caregivers, it is pertinent to question

DOI: 10.1201/9781003475361-5

whether all cases of distress constitute "burnout" or if additional contributing factors merit exploration.

At present, "burnout" is frequently applied as an umbrella term encompassing holistic exhaustion, distress, and frustration across professions. As with any process influencing patient wellness, examining root causes is crucial to effectively address symptoms and develop targeted interventions that sustainably enhance wellbeing. The initial step in this process involves clarifying and refining the terminology used to describe the experiences of caregivers.

Mental Health Matters

> "The high-stakes issue of mental health is growing as a concern among employees, and normalizing the conversation around it is critical to improving and addressing mental health problems like burnout, depression, anxiety, addiction, and the disengagement that results from these problems."[1]
>
> – Fisher and Phillips (2021)

Definition: Burnout is a multifactorial, globally recognized occupational syndrome resulting from chronic workplace stress that has not been successfully managed.

The **World Health Organization** officially classified burnout as a medical diagnosis, including the condition in the International Classification of Diseases:

> "A syndrome conceptualized as resulting from chronic workplace stress that has not been successfully managed."[2]
>
> – World Health Organization (2019)

Burnout is associated with 4 symptoms:

1. Feelings of energy depletion, exhaustion, and fatigue
2. Increased mental distance from your job
3. Feelings of negativism or cynicism related to your job
4. Reduced professional efficacy[2]

Why It Matters: Burnout is more than exhaustion – it affects organizational stability, patient care, and workforce sustainability. Recognizing its many layers allows for both compassionate understanding and actionable intervention.

Stress and burnout are not the same. The fatigue associated with burnout is different than the stress you have from having worked long shifts, but it can feel similar. Importantly, the fatigue layers and linger. As it is multifactorial and has inputs that are outside of the individual worker's control, addressing the sources of fatigue can be challenging. Burnout results from cumulative stress, which makes management techniques that may help address some specific work stressors less effective.

That said, the research from Gallup indicates that working less is not enough to prevent burnout. Employee burnout is strongly influenced by how they experience their workload and work-related stressors. Primary causes found in this survey that contribute to worker burnout: unfair treatment, an unmanageable workload, lack of role clarity, and lack of communication and support from a manager. If the work being done is fulfilling, meaningful, and recognized, the job-related incidence of burnout is significantly decreased.[3]

Jennifer Moss has researched and published widely on workplace burnout. Some of her better-known works are *Beyond Burned Out* (2021), *The Burnout Epidemic* (2021), and *Unlocking Happiness at Work* (2025). Dr. Moss states that:

> This language acknowledged that burnout is more than just an employee problem; it's an *organizational* problem that requires an organizational solution. . . .
>
> While these are all organizational issues, we still prescribe *self-care* as the cure for burnout. We've put the burden of solving the problem squarely on the shoulders of individual employees. "Let's just recommend more yoga, wellness tech, meditation apps, and subsidized gym memberships – that'll fix it," we say. But those are tools for improving well-being. When it comes to preventing burnout specifically, they won't be effective. We desperately need upstream interventions, not downstream tactics.[4]

Six Main Causes for Occupational Burnout from Maslach et al.

1. Unsustainable workload
2. Perceived lack of control
3. Insufficient rewards for effort
4. Lack of supportive community
5. Lack of fairness
6. Mismatched values and skills[5]

Impactful Strategies to Address Occupational Burnout

These burnout "antidotes" are preventative measures taken by both the employee and by the leadership team. The *Wellbeing at Work* authors offer 5 prevention action items:

1. Make sure everyone in your organization knows their strengths. Use a strengths-based strategy to design an employee experience – from attraction to hiring to onboarding, engagement, and performance – that leads to a culture of high development (as well as satisfaction and productivity!).
2. Remove abusive managers. No organization should tolerate managers who destroy the lives of people you rely on to get the work done. In the workforce of today, bad managers are your highest risk.
3. Up-skill managers to move from "boss" to "coach." Use proven methods to transition your managers' mentality by focusing on a long-term professional development path that starts with learning about high-performance teams. Each manager should become an expert at setting goals and providing meaningful feedback regularly (frequency determined by size of team being managed, by how feedback is preferred to be given by the worker, and by ensuring personal and professional goals can be achieved while aligning with the higher organization or business itself).
4. Make wellbeing a part of career development conversations. Once managers establish trust, they, along with their teams, can dream big together – not just about career goals and development but about life and overall purpose and wellbeing.
5. Working less does not mean happier work. It is important to find ways to make professional wellbeing a focus in your organization (as a value!)[6]

Factors That Predict Lower Levels of Burnout

Research Effort with Moss, Leiter, Maslach, and Whiteside

A survey of 1,500 respondents in 46 countries in various sectors, roles, and seniority levels was done in 2020, with 67% of respondents working at or above a supervisor level. The following key factors were found to be associated with lower levels of burnout:

1. Feeling a sense of purpose. In this research study, burnout scores declined as purpose scores increased.

2. Having a manageable workload (including clear communication of priorities, providing adequate resources, and encouraging inter-team collaboration and delegation when appropriate).
3. Feeling that mental health can be discussed at work. This requires a culture of psychological safety, which Amy Edmondson, an expert on the topic, defines as "a climate in which people are comfortable being (and expressing) themselves." [6]
4. Employees having access to mental health support.
   a. Mental health resources clearly communicated to team
   b. Peer-to-peer support (having interested team members be trained in "mental health 101" and be prepared to bridge to appropriate mental health resources and support when needed)
   c. Have a crisis debriefing plan and strategy
5. Having an empathic manager.
   a. Communicating empathically increases job satisfaction, reduces burnout, and is highly correlated with enhanced wellbeing.[7]
   b. Empathic leadership requires 3 things:
      i. Acknowledging and overcoming any personal biases or privileges
      ii. Listening to learn (active listening)
      iii. Taking actions informed by collaborative, inclusive decision-making
6. Having a strong sense of connection to your community (friends, family, colleagues).[8]

Key Sources and Perspectives

- WHO defines burnout as a syndrome tied specifically to workplace stress.
- Christina Maslach identifies six root causes: workload, control, reward, community, fairness, and values alignment.
- Jennifer Moss emphasizes system-level change over individual self-care.
- Gallup's 2021 study links burnout to poor leadership, unclear expectations, and lack of recognition.

Workbook Prompt

Reflect on your current workplace stressors. Which of these stressors align with the Maslach six dimensions? What are current strategies that exist or that may need to be created to address these stressors in your professional ecosystem?

Quote Box

"Burnout is not an individual failure, it's a workplace issue that demands a workplace solution."

– Jennifer Moss

Section 2: Paradigm Shifts in Caregiving Professions

With the high incidence of discussions about workplace burnout, individuals considering or already working in caregiving roles may view burnout as a likely outcome. Caregiving work is indeed challenging, but burnout is not an inevitable terminus. Developing awareness, implementing strategies, and establishing cultural support can help caregiving professionals and leaders effectively manage workplace stressors. This raises the question: What are the objectives of occupational wellness?

> Wellness is "a state of complete physical, mental, and social well-being and not merely the absence of disease or infirmity."
>
> – World Health Organization[9]

Occupational Wellness

Definition: Creating personal satisfaction and enrichment from one's work via:

- Preparing for and entering work that is consistent with one's personal interests and values
- Gaining satisfaction from work that is personally enriching and rewarding
- Expanding and evolving one's skills and interests throughout life

Occupational wellness refers to maintaining a balance between work and leisure time, forming professional relationships with colleagues, and managing workplace stress. Identifying meaningful and financially sustainable employment contributes to occupational wellness. The workplace culture, including how stressors are managed, can influence individual and team experiences of holistic wellness.[10]

Common Workplace Stressors

- Interpersonal conflict
- Communication problems

- Gossip
- Bullying
- Harassment
- Discrimination
- Low motivation and job satisfaction
- Performance issues
- Poor job fit

For veterinary professionals, all of these stressors, if not addressed, can lead to decreased performance and productivity, decreased job satisfaction, and a wide variety of mental and physical health concerns. For medical care providers, additional contributors to the experience of exhaustion and distress include moral distress, empathic strain, secondary traumatic stress, emotional labor, and physical exhaustion.

These combined with the feelings of not feeling valued, appreciated, or appropriately financially compensated (vet team > veterinarians) contribute to veterinary professionals feeling less engaged and energized by the work that they do. These pieces combined with many veterinary professionals not having a clear self-care strategy in place and stress management techniques while at work and between shifts can contribute to more chronic accrual of the elements that combine to the "exhausted" pillar of workplace burnout.

Practice Evolution

A shift from reactive to preventative approaches is necessary to normalize workplace wellbeing. Overcoming cultural barriers to self-care – especially in fields that value competence and perfectionism – is essential. Fostering a positive clinic culture enhances mental health, increases engagement, and reduces staff turnover. In human medicine, increased investment is being made in clinical wellbeing as a quality and safety imperative. And the fields of education, law, and social services are adopting trauma-informed leadership models.

From the Merck Wellbeing Study IV, the top 4 elements that were found to contribute to professional satisfaction included:

- Strong sense of belonging to the team
- Sufficient time is allotted for high-quality patient care
- High degree of trust in the organization
- Candid and open communication among team members[11]

In all of these professional fields, increased resources that support physical and mental health are being provided. Key to utilization is communicating what resources are available and how to access them.

Box Out

"The stigma around acknowledging mental health needs remains a significant barrier – normalizing conversations is the first step to systemic reform."[12]

– *Work Better Together*

Strategies to Increase a Culture of Trust, Appreciation, and Occupational Wellness

Individual Strategies: These are daily personal actions – like mindfulness, setting boundaries, or therapy – that support mental and physical health and build resilience.

Workplace Strategies: Organizations can promote wellbeing through flexible schedules, open communication, mental health resources, training on resilience or burnout, and creating a supportive environment.

Industry Strategies: At the systemic level, initiatives include national mental health programs, legislative advocacy for better work standards, and cross-organization collaborations to benefit the profession as a whole.

Workbook Prompts

- What industry-level wellbeing initiatives have you encountered or participated in? How effective were they in addressing the challenges they aimed to solve?
- What attitudes about caregiving or professionalism have you internalized? Which ones may no longer serve you or your team?
- Are there any specific resources or programs you have found helpful, such as wellbeing workshops, partnerships with mental health organizations, or peer support networks?
- What gaps do you see in the current wellbeing strategies at the industry level, and what solutions might help fill these gaps?
- Can individualized strategies at a large scale create systemic change? Why or why not? (For example: Do courses that teach stress management skills to vet students and new veterinary graduates at an individual level have the potential to create large-scale change?)
- In what ways do you see industry-wide wellbeing strategies impacting the future of veterinary work culture?

Section 3: International Trends in Workplace Wellness

Adopting both a OneHealth and a multidisciplinary approach helps expand understanding and improve workplace wellbeing discussions. This section offers key research and findings from outside veterinary medicine to encourage curiosity and support effective strategy development for veterinary organizations.

Global Data Highlights

- The Wellable's 2025 Employee Wellness Industry Trends Report shows that organizations are moving from fitness and nutrition programs to mental health, resilience, and social connections.[13]
- Sapien Labs' 2024 study shows employees with strong social connections report double the mental wellbeing of isolated colleagues.[14]
- The World Economic Forum lists mental health and wellbeing as top workplace priorities for the decade.

World Economic Forum – Future of Jobs 2025

- **Employee health and wellbeing** have emerged as a critical business priority, with 64% of employers identifying it as a key talent attraction and retention strategy – indicating a significant shift toward prioritizing mental health in the workplace.
- **Resilience, flexibility, and agility** rank among the most in-demand core skills that employers are seeking, suggesting that emotional intelligence and adaptive capabilities are becoming central to workplace success.
- **The care economy**, including counseling and social work roles, is projected for substantial growth, while skills in self-awareness and motivation are increasingly valued, pointing to expanding opportunities in mental health and personal development services.[15]

Box Out: Important Consideration in Veterinary Environments

It's one thing to advertise a culture of health and wellbeing for the veterinary team as a priority; it's another for each associate (from kennel tech to senior leader) to live and embody this as part of a shared healthy culture.

The *Mind the Workplace 2024* report by Mental Health America highlights the ongoing mental health challenges in U.S. workplaces, with work-related stress impacting sleep and personal relationships for many employees. Gen Z and Millennial workers report high stress due to limited support and autonomy, while a lack of psychological safety and trust contributes to higher turnover, burnout, and disengagement.

The report reveals that only half of employees feel their workplaces prioritize mental health, and those with supportive managers and clear communication report better mental health and productivity.

Companies that excel in employee mental health foster trust, transparent communication, and flexible work practices, which improve engagement and reduce turnover. Mental Health America urges organizations to invest in both individual and organizational strategies that prioritize mental well-being to create thriving, productive work environments.[16]

Key Points

- A significant number of employees report that their workplaces do not adequately support mental health, which contributes to burnout and increased turnover.
- Work-related stress continues to affect employees significantly, with impacts on sleep and personal relationships, particularly for Gen Z and Millennials.
- Organizations with strong mental health initiatives, including manager support and work-life balance adjustments, tend to see higher productivity, better employee retention, and an overall more positive work environment.

HBR Article on Workplace Wellbeing Programs

Workplace wellness programs are widespread among U.S. employers (85%) and projected to reach $94.6 billion globally by 2026, but they often fail to improve employee wellbeing due to a focus on individual-level interventions, such as wellness apps and mindfulness, rather than addressing systemic factors like workload and management practices. This can lead to "carewashing," where programs seem superficial and fail to engage employees.

To be more effective, wellness programs should take a holistic approach, combining individual and organizational changes, such as promoting work-life balance, tracking mental health outcomes, and providing leadership training. Aligning with global standards like ISO 45003 and prioritizing transparency can help create a supportive culture where mental health is integral, improving both employee satisfaction and program impact.[17]

Key Points

- Workplace wellness programs often focus on individual interventions (like wellness apps and stress management training)

but fail to address systemic issues such as high workload and poor management practices. Such individual interventions are often seen as superficial and disingenuous by employees.
- Shifting focus to organizational changes, like workload management and mental health training for leaders, can more effectively support employee wellbeing. This supports a systems approach that supports everyone in the ecosystem with shared responsibility and accountability.

By combining leadership support, flexible work policies, and structured well-being metrics, organizations can better address employee needs and improve overall program effectiveness.

Sapien Labs Work Culture and Mental Wellbeing

A comprehensive study of 54,831 employed individuals across 65 countries reveals that workplace relationships and sense of purpose are the primary drivers of employee mental wellbeing. The Sapien Labs 2024 report finds that good relationships with colleagues and feeling pride and purpose in work have nearly twice the impact on mental health as traditional work-life balance factors like workload and schedule flexibility.

The findings challenge the traditional concept of work-life balance, suggesting that meaningful connections and purpose at work are more crucial for mental wellbeing than reduced workloads or increased flexibility. The report emphasizes that humans are social creatures who need strong relationships and meaningful work to thrive, even if it means working more or with less flexibility.

Rather than focusing primarily on individual wellness interventions like mindfulness apps or therapy, the report suggests organizations should prioritize building effective teams with a clear sense of purpose and manageable workloads. This reflects the understanding that work life is as personal as home life, and the relationships and meaning we find (or fail to find) at work significantly impact our overall mental wellbeing.[14]

Key Points

- A positive work culture that includes a sense of belonging, clear communication, work-life balance, role clarity, and supportive management is crucial for good mental health among employees.
- Negative workplace factors, such as lack of support and poor communication, significantly increase rates of anxiety, depression, and burnout, especially among younger employees in high-stress roles.

- To improve mental wellbeing and reduce turnover, organizations should foster inclusive cultures, support mental health openly, and train leaders to adopt empathetic, health-focused management practices.

Strategies to Support Veterinary Workplace Wellness

- Prioritize relationship-building and trust when creating SOPs, in team conversations, and during professional development initiatives.
- Design inclusive programs rooted in shared values and cultural awareness.
- At team meetings, ask colleagues "What is going well? What is working? What does "even better yet" look like for you, for the team, for the organization?

Workbook Prompts

- What parts of your workday feel most connected to meaning, autonomy, or relationship?
- Where do you feel isolated or depleted?
- What experiences, interactions, and contributions energize you?

Box Out

"Employees with strong relationships and a sense of purpose report nearly double the mental wellbeing of those relying only on work-life balance."[14]

– Sapien Labs, 2024

Section 4: Systems vs. Symptoms – Reframing Wellness Programs

The importance of communities and cultures of support for the adoption, practice, and maintenance of wellbeing practices cannot be understated for caregiving communities to have increased health and decreased workplace burnout.

Recognizing the link between workplace culture and team wellbeing is crucial for the veterinary profession. While individual wellbeing practices may offer some short-term relief, long-lasting improvement depends on addressing underlying causes through systemic change.

Sustainable solutions require organizational and cultural transformation to prevent compassion fatigue proactively. Stakeholders should pursue transformative changes and support further research into effective wellbeing practices. Collaborative, "person-first" approaches can drive these necessary changes, leading to healthier, more sustainable careers for veterinary teams.

Personal responsibility is important AND needs to be nourished and fostered by an environment/ecosystem, culture, and leadership that supports safety, collaboration, engagement, and innovation. All of these elements contribute to holistic wellness, professional satisfaction, and resiliency in the individuals, teams, and practices/businesses. Community accountability is paramount.

Critique of the Individual-Focused Approach

- Wellness initiatives that focus only on the individual (e.g., meditation apps, yoga breaks) without addressing workload, leadership, or values misalignment can become a form of "care-washing" and blame-shifting.
- True change demands that leaders look upstream and address structural drivers of distress.

Effective Approaches

- ISO 45003: An international standard for psychological health and safety at work. This standard is applicable to workplaces and organizations of all sizes and sectors to support the development, implementation, maintenance, and ongoing improvement of healthy and safe workplaces[18] (https://www.iso.org/standard/64283.html).
- Whole-system audits that include workload, communication, autonomy, and purpose.
- Building emotional culture as well as cognitive culture at all levels of the organization.

Reflection

How does your workplace currently address the root causes of stress, disengagement, or turnover?

Section 5: The Veterinary Profession – Where Are We Now?

Traditionally, both human and veterinary medical caregiving have upheld a work culture that values stoicism and self-reliance. The ability to compartmentalize challenging emotions and defer personal needs has often been regarded as an essential trait for effective clinicians or nurses. Unfortunately, fostering healthy boundaries for self-care – including fundamental needs such as nutrition and sleep – has sometimes been stigmatized, perceived as a sign of inadequacy within the profession. Persistently prioritizing the needs of others over their own can result in significant health and business consequences.

When combined with a lack of safety or support for discussing the mental and physical demands inherent to caregiving, this "other-focused" approach may contribute to isolation, increased health issues, compassion fatigue, and burnout among medical professionals.

Support systems for veterinary caregivers' wellbeing are developing out of necessity, as attrition rates reach unprecedented levels. Historically, responsibility for managing occupational stress and mental health has rested primarily on individual caregivers. However, recent shifts in veterinary training and clinical mentorship now place greater emphasis on psychosocial skills and human factors as integral components of clinical preparation.

Despite these advances, stigma around acknowledging the need for support remains a considerable barrier to seeking help and investing in required resources. Working conditions highlight the urgent need to prioritize the mental, emotional, and physical health of the veterinary workforce. A 2021 evidence-based review emphasized making health care worker wellbeing a pressing public health priority.[19]

Current Veterinary Research

- Merck's 2020 Veterinary Wellbeing Study reported declining mental health compared to 2017 data, despite increased awareness.
- Merck's 2023 Veterinary Wellbeing Study expanded its scope to examine the mental health and wellbeing of veterinary team members – including veterinary technicians and managers – in addition to surveying veterinary clinicians. It underscored a positive trend in veterinary practices and professionals taking a more proactive approach to mental health.
- Steffey et al. (2023) found a strong correlation between toxic leadership and turnover intent.
- John Neill estimates burnout-related attrition costs over $1 billion annually in lost productivity and retraining.

Merck Veterinary Wellbeing Study II (2020)

On average, veterinarians experienced higher rates of burnout than did physicians, despite working substantially fewer hours per week. Suicide ideation and planning were more prevalent among veterinarians than non-veterinarians, and veterinarians were 2.7 times as likely to attempt suicide as non-veterinarians.

Alcohol use, including binge drinking, among veterinarians was consistent with that found in the adult general population. Recreational drug use was uncommon among veterinarians; cannabis use by veterinarians was significantly lower than among non-veterinarians.

The factors having the strongest associations with burnout scores on multiple regression analysis were lack of work-life balance, not enjoying work, not finding work invigorating, and having personal conflict with 1 or more work associates.

Stress in veterinarians was considered the most critically important issue facing the profession by 92% of respondents. Financial stress in the form of high student debt and low incomes and poor work-life balance were major contributors.

Box Out: Specific Recommendations by the Merck 2020 Studies' Authors

Bring Distress and Mental Health Front and Center

Although most respondents agreed with or had a neutral response to statements about mental health and treatment, the stigma associated with these factors unquestionably remains. Such stigma can be reduced if mental health is addressed during in-hospital meetings and other gatherings of veterinarians.

During team meetings, acknowledge that veterinary medicine is a stressful profession and that feelings of burnout are common. Group discussion will demonstrate to those feeling overwhelmed that they are not alone and give them permission to acknowledge their feelings.

Periodically inviting local mental health professionals to address team meetings will also make local resources more familiar to team members. Associate veterinarians and other employees could be given permission to schedule meetings with counselors – even during work hours – when needed. In addition, several tele-behavioral health resources are now available that individuals can access on their own time (e.g., E-counseling.com).

Create A Stress Management Plan

Given that stress is endemic to veterinary medicine and other health professions, it is important for each individual to have a stress management plan. Indeed, veterinarians with such a plan in the MAH Veterinarian Wellbeing Study II were less likely to experience serious psychological distress and more likely to be flourishing. Online resources are available that provide techniques for dealing with stress (e.g., vetwellbeing.com and avma.org/resources-tools/wellbeing).

Balance Work and Healthy Activities

The value of a good balance between work and healthy nonwork activities was repeatedly supported by the study findings. Good work-life balance contributes to higher wellbeing, less serious psychological distress, less burnout, and higher job satisfaction. As also identified in the 2017 study, spending time with family and friends were among the nonwork activities with the greatest positive impact. Other activities included exercising, having a hobby, traveling for pleasure, and reading for pleasure.

Engage A Financial Planner

Given the prevalence of burdensome student debt and the huge impact of financial stress on veterinarians, it is prudent to seek professional financial advice. Veterinarians who worked with a financial planner were only a third as likely to experience serious psychological distress and had significantly higher levels of wellbeing, compared with those without a financial planner.

For those with student debt, the VIN (Veterinary Information Network) Foundation Student Debt Center provides many resources (vinfoundation.org/resources/student-debt-center). Other resources are also available (e.g., avma.org/membership/SAVMA/financingyour-veterinary-medical-education and vmae.org/veterinary-debt-initiative).

Fee-based financial planners are advisors who charge only for their professional services and are not affiliated with firms whose goal is to sell investment vehicles or other services. Fee-based financial planners can be found through the National Association of Personal Financial Advisers (NAFPA.org). For those who cannot afford a financial adviser, pro bono resources are available from the NAPFA Foundation (napfafoundation.org), Financial Planning Association (onefpa.org/advocacy/Pages/Pro_Bono_Financial_Planning.aspx), and National Foundation for Credit Counseling (nfcc.org).

Limit Time on Social Media

Spending more than an hour a day on social media was negatively associated with wellbeing and mental health in both the 2017 and 2019 studies. Also, those who spent less time on social media had less exposure to cyberbullying or vicious reviews.[20]

Note from Sonja

Many organizations have Employee Assistance Programs that also include financial advising as a free or discounted benefit.

Merck Wellbeing Study IV, 2023 – Key Findings on Veterinary Occupational Wellness

Who responded to the survey: 4,700 DVM respondents, 2,300 staff respondents; DVM – 4 generations, predominantly white, female > male; Team – 4 generations with younger folks; white; female > male.

Many results were found to be interesting, particularly to see the trends compared to the prior 3 years of the Merck Wellbeing Studies. Some of the highlights from this most recent survey include:

- Career satisfaction – DVMs 75% somewhat/extremely satisfied; "others" thought 43% were satisfied.
- Media and trainings lean toward negative concepts (mental health, burnout).
- The reality of professional satisfaction is more positive than the general perception of respondents. This was a surprising and important finding on perception vs reality and how both media and in-practice conversations around "wellness" and "burnout" impact veterinary workers' views of the profession.
- Higher job satisfaction with DVMs (especially when compared to the U.S. general population).
- Work schedules are improving: 40 hours average in 2023 compared to 45 hours in 2020.
- Retirement is the likely key reason for leaving vet profession (age > 65 yrs).
- Veterinarian wellbeing – 53% have a high level of wellbeing and flourishing; approximately 30% are "getting by."
- More professional experience means higher wellbeing (same trend in the U.S. population; especially for individuals > 45 years old).
- Burnout in veterinarians is similar to general population.
- Veterinarians experience higher levels of the *exhaustion element* of burnout.
- Vet teams are more likely to experience burnout and exhaustion than the general population.
- Managers and DVMs tend to have less burnout than the rest of the team: DVM: 18%, Managers/Admin 20%, Vet team 28%.[12]

The Ripple Effect of Burnout: From Individual Strain to Organizational Impact

In 2023, Steffey et.al. published two papers with complimentary content that took a deep dive into what is known to date for caregiving professionals in general and veterinary workers on the demographics, the individual effects, and the organizational impacts of workplace stress and burnout. These articles highlight preventable causes of occupational suffering in the veterinary profession and introduce workplace management and culture changes as solutions to address burnout and improve wellbeing.

Burnout in veterinary medicine is more than personal fatigue – it is a systemic challenge with serious implications for clinician health, team culture, and organizational sustainability. Recent global research makes it clear: Chronic stress and burnout affect every level of veterinary practice.

I personally have referenced these articles dozens of times since they were written. In addition to the well-organized and clearly presented content, each publication also has several hundred citations providing excellent additional associated content on occupational distress, wellbeing, and burnout. Some of the salient highlights from these two exceptional articles are shared here in the context of this chapter.

Understanding Burnout in Veterinary Medicine

Burnout in the veterinary profession is both pervasive and consequential, affecting individuals, organizations, and the quality of patient care. Recent literature highlights alarming statistics:

- 92% of veterinarians identify stress as the most critical issue in their profession.
- 43% often feel distressed or anxious about their work.
- 40% are considering leaving the profession due to poor work-life balance and mental health challenges.
- Burnout rates range from 27% to 89%, nearly 40% higher than comparable physicians.
- Women veterinarians and those with less experience report higher burnout risk, and veterinary students and early-career clinicians face particularly high vulnerability.

Burnout is recognized by the ICD-11 as an occupational phenomenon, not a mental illness, and is characterized by emotional exhaustion, cynicism, and reduced professional efficacy. Three main subtypes are described:

- Frenetic: driven by work overload
- Underchallenged: caused by monotonous tasks
- Worn-out: stemming from feelings of hopelessness

The Pathophysiology and Health Effects of Burnout

Burnout is not only a psychological experience; it has biological underpinnings that affect long-term health:

- Chronic stress activates the sympathoadrenal and hypothalamic-pituitary-adrenal (HPA) systems, creating hormonal imbalances and chronic low-grade inflammation.
- Functional brain changes: Disrupted connectivity between the amygdala and prefrontal cortex impairs emotional regulation.
- Increased mortality risk: Each unit increase in burnout score correlates with a 35% increased mortality risk.
- Associated with cardiovascular disease, metabolic disorders, sleep loss, and cognitive deficits such as impaired memory and attention.
- Secondary conditions include depression, anxiety, and suicidal ideation.

The Organizational Toll: Culture, Team, and Financial Loss

Burnout does not remain confined to the individual – it ripples through veterinary organizations, influencing team cohesion, workplace climate, and financial stability.

Impacts on Organizations

- **Absenteeism and Turnover:** Burnout drives high turnover, with the cost of replacing a veterinarian estimated at $104,000, and technicians at $59,000 per person.
- **Financial Losses:** Burnout in U.S. veterinary practices results in an estimated $2 billion in lost revenue annually (see the next subsection).
- **Productivity and Errors:** Burnout correlates with increased medical errors – clinicians experiencing burnout have twice the risk of patient safety incidents.

Impacts on Team Climate and Workplace Culture

- Burnout contributes to toxic work environments, decreased morale, and breakdowns in teamwork.
- Poor work-life integration fosters a culture of presenteeism and pathological altruism, reinforcing unhealthy norms.
- Teams in supportive work climates with work-life balance policies report lower burnout and better collaboration.

Impacts on Patient Care

- Each 1-point increase in depersonalization correlates with an 11% rise in error likelihood.
- Suboptimal care is reported 53% of the time by burnt-out clinicians, compared to 21% of peers.[21]

Solutions That Work: Systemic and Individual Strategies

Research is clear: Burnout is not solved by resilience alone. Individual strategies matter, but without systemic change, they fail to address root causes.

Organizational-Level Solutions (Most Effective)

- **Redesign workload:** Even a 10% reduction in task load lowers burnout odds by 33%.
- **Promote work-life integration:** Policies and leadership that normalize self-care and flexible scheduling.
- **Create psychological safety:** Encourage open discussion of mental health without stigma or fear.
- **Address cultural norms:** Actively challenge beliefs that personal health should be sacrificed for patient care.
- **Apply the Hierarchy of Controls:** Focus on systemic changes (workload, staffing, culture) before individual training.[21, 22]

Individual Supports (within Supportive Systems)

- Mindfulness and stress management practices
- Access to counseling or peer support programs
- Protected time for rest and recovery – free from guilt or reprisal

Key Takeaways

- Burnout harms both people and organizations. It erodes wellbeing, increases errors, and drains financial resources.
- Systemic fixes (culture, workload, leadership) matter more and have more sustainable impacts than individual coping strategies alone.
- Organizational commitment to clinician health fosters stronger teams, better care, and sustainable veterinary practice.

Economic Cost of Burnout in Veterinary Medicine

Burnout negatively impacts the veterinary profession's economic health, affecting both supply and demand for services through factors like decreased consumer confidence, perceived care quality, employee turnover, reduced

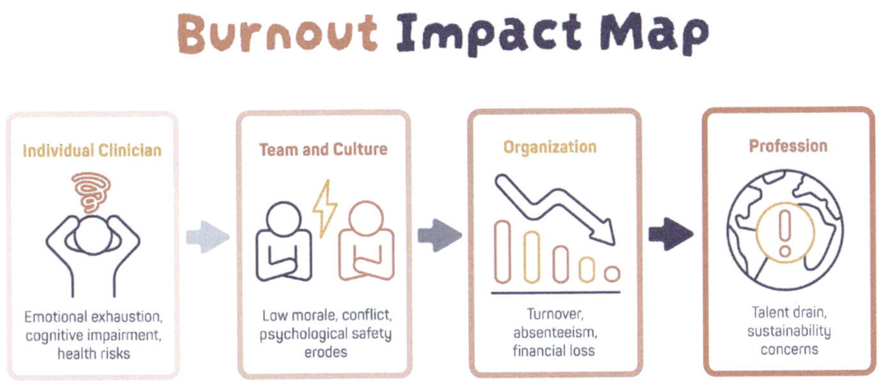

Figure 5.1 Burnout impact map.

work hours, and professionals leaving the field. Burnout can also hinder future recruitment.

Dr. Neill's team calculated burnout rates and examined how it influences turnover and reduced work hours. They considered turnover costs such as recruitment, training, relocation, bonuses, and lost revenue from staffing gaps or fewer working hours.

The average total industry cost is $0.928 billion to $1.005 billion annually in lost revenue because of burnout among veterinarians. That amounts to about 2.1% of the total industry revenue for 2020. Put another way, that amount could pay for the class of 2020's education two times over.

If veterinary technicians' burnout is added to the equation, economic losses increase to about $1.93 billion annually. Dr. Neill said, "There's a lot of techs out there, and burnout is also relatively high among them."

Dr. Neill suggested the following action items to address burnout at the practice level:

- Hold monthly meetings to focus on work-life challenges and difficult issues in management of patient care.
- Improve workflow by offloading nonessential tasks to non-veterinarian staff members.
- Remove bottlenecks in patient rooms with regard to medication reconciliation, vaccinations, and data entry.
- Reduce time pressure with plans for future increases in visit time for primary care.
- Institute a new prescription phone line to free up veterinary technicians if the practice has a pharmacy in house or other ways to distribute food or flea and tick medications.
- Present work-life data as a platform to discuss issues within the clinic.[23]

Box Out

> "If they implement the right interventions, managers can make a larger difference to alleviate burnout in their organizations than individuals can do on their own."
>
> – Dr. Sunita Sah, associate professor of management and organizations at the Cornell SC Johnson College of Business

Implications

- Burnout prevention is both an ethical imperative and a business strategy.
- Career satisfaction is directly tied to values alignment, autonomy, and support from leadership.

Workbook Prompt

What changes would most significantly reduce stress or increase energy in your current work life?

From Reactive to Proactive: A New Wellness Paradigm

The personal wellbeing of each veterinary professional has broader implications for veterinary teams and organizational resilience. Healthier and better-supported individuals foster stronger, more effective teams, thereby promoting high-quality patient care and enhancing overall practice sustainability.

These considerations play a critical role in discussions about the long-term viability and attractiveness of veterinary medicine as a career. It is an ethical imperative for veterinary schools, organizations, and leaders to devise strategies that address and alleviate occupational stressors prevalent in veterinary medicine and caregiving at large.

For systemic reform in resource development, promotion, and implementation, an improved understanding of the distress experienced by veterinary caregivers must be recognized as a valid target for investment. The overarching challenge is to move beyond stress management and prevent burnout rather than just reacting to it when it occurs for individuals and teams.

Guiding Questions

- How do we assess wellness in our teams?
- What signals indicate occupational wellness or distress?

- How can we support staff across career stages and roles?
- What systems or policies in your organization most influence occupational wellness

Highlights

- Importance of modeling self-care and boundary-setting by all levels of leadership
- Essential to have "holistic wellness" as a team and practice value supporting

Section 6: Effective Support for Organizational Wellbeing and Leadership Transformation

Evidence-Based Prevention Strategies

- Hire and promote based on emotional intelligence and coaching ability, not just technical skills.
- Provide ongoing professional development that includes mental wellbeing and team dynamics.
- Encourage micro-breaks and team debriefs after emotionally intense cases.
- Normalize asking for help and having recovery rituals.
- Implement Emotional Intelligence and Psychological Safety training for leadership and mentors.

Toolkits and Approaches

Note: Details available on these efforts in Chapter 7's *Resources:*

- DARN/DARNCAT for motivational dialogue.[24]
- Mental Health First Aid training – mental health crisis intervention and suicide prevention approaches for all leaders and staff.[25]
- PERMA+4 framework for culture assessment (available through the FVMA).
- Psychological safety standards from AAHA/Veterinary Visionaries.
- WSAVA Veterinary Wellbeing Guidelines (2024).
- AVA workforce survey (2018) to inform strategic changes.
- Cultivating Safe Teams AWARE Program for psychological health and safety awareness. (AVA)

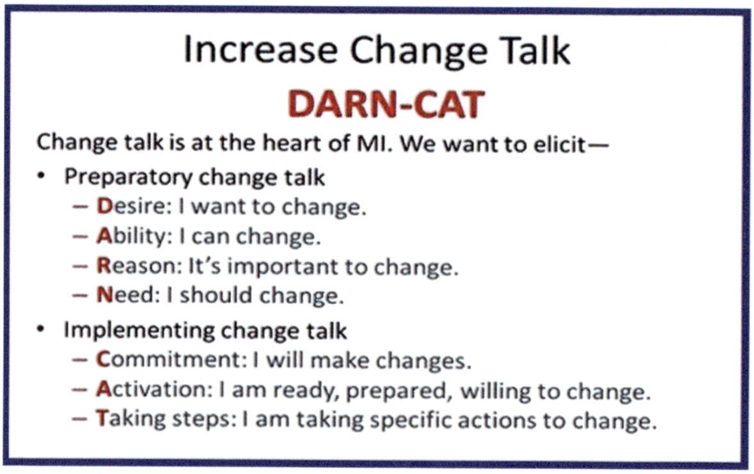

Figure 5.2 Increase Change Talk: DARN-CAT
Source: Miller and Rollnick, Motivational Interviewing: Helping People With Change, 3rd Edition.

- ◆ Graduate mentoring programs and MHFA crisis support (MentorVet; Lincoln Institute).
- ◆ RCVC Mind Matters (2025), VetLife (2022), and Thrive's Prevent–Protect–Promote resources (2024).

Box Out: Veterinary Leader's Sharing

"We have been trying to move the dial on more creative scheduling for ER doctors (i.e. 4 × 10-hour shifts instead of 3 × 12-hour shifts). When having these conversations, there seems to be an invisible barrier of 'this is how we've always done things', or 'Don't change a system that works, just find the people who are willing.' People in the veterinary field are brilliant problem solvers. If leaders would hand the problem they are needing to solve over and share parameters around what is/is not negotiable, the team likely would come up with a creative solution themselves."[26]

– This was a personal communication from a veterinary colleague and leader in 2025

Leadership and Cultural Transformation

- ◆ Balance accountability with individual growth to strengthen team wellbeing.
- ◆ Facilitate conversations around values and team agreements.
- ◆ Empower peer-to-peer boundary setting.
- ◆ Lead systemic redesign rather than focusing solely on resilience.

Quotes and Expert Commentary

- "We are treating burnout all wrong. It's not about resilience; it's about redesigning."[27] – Dana Varble
- "Empathic managers reduce burnout, increase job satisfaction."[28] – Helen Riess

Practice Spotlight

- Galaxy Vets' discussion on the inclusion of mental health in the veterinary oath (as exists currently in the human physician oath).[29]
- Initiatives for cultural support and change (please see Chapter 7 for more details):
  - Not One More Vet (NOMV) – supportive community and CLEAR Blueprint
  - World Small Animal Veterinary Association (WSAVA) – guidelines and CE
  - American Veterinary Medical Association (AVMA) – wellbeing and professional development trainings and initiatives
  - American Animal Hospital Association (AAHA) and Veterinary Visionaries
  - Australia Veterinary Association (AVA) and Thrive
  - British Veterinary Association (BVA)/Royal College of Veterinary Medicine and Mind Matters
  - Canadian Veterinary Medical Association (CVMA) and The Working Mind
  - International Association of Veterinary Social Work (IAVSW)

Workbook Prompts

- Where is your organization already succeeding in supporting wellbeing?
- What's one small change that could make a big impact?
- Who in your workplace models the kind of leadership you want to follow? What are they doing differently?
- What is one belief you want to shift in your team culture?

Chapter Summary

Cultivating occupational wellness and diminishing workplace burnout are multifaceted challenges that demand collective insight and coordinated action. Global research reminds us that individual resilience alone is not enough – we must reimagine systems, workplaces, and expectations. The veterinary profession stands at a crossroads, with an opportunity to lead not only high-quality care for animals but also in much-deserved meaningful care for veterinary caregivers.

Knowing what you now know, you have the power to make changes within your practices. You can advocate for yourself and others. We talk a lot about systemic change, and this is how it happens. Collectively, we are coming together and advocating for a healthier profession

Key Takeaways

- Burnout is a systemic issue, not a personal failing.
- Global insights point to relationships, autonomy, and values as core drivers of wellbeing.
- Veterinary-specific research highlights urgent needs and hopeful innovations.
- Sustainable solutions require leadership, empathy, and strategic redesign.

Box Out

"Burnout may be real, but it doesn't have to be the reality for you or your team members."[30]

– Dr. Quincy Hawley, GetMotiVETed University

Bridge to Chapters 6 and 7: Future-Forward Focus on Designing Sustainable Veterinary Careers and a Thriving Veterinary Profession

Chapter 6 moves us from diagnosis to design. How do we build veterinary careers – and entire systems – that prioritize purpose, flexibility, and flourishing? Through case studies, strategic frameworks, and visionary leadership, researchers and those committed to culture change have explored what it looks like to make wellbeing central to how we train, support, and evolve in veterinary medicine.

Let's move from intention to transformation. Tools, resources, and supportive community are needed!

Chapter 6 builds upon what was offered thus far by gathering psychometric tools, assessments for individuals and teams, and training/certification opportunities. We then reach Chapter 7 as our final stop on this journey

Figure 5.3 The Wellness Journey – Sustainable through Community and Shared Purpose.

of reflection, growth, and strategic fortification. This chapter is a true celebration. I have crowd-sourced a curated collection of impactful and interesting books, blogs, articles, podcasts, veterinary initiatives, and supportive communities to support further engagement and growth, as the learning journey is never done!

References

1. Fisher, J, and Phillps, A. (2021). *Work Better Together: How to Cultivate Strong Relationships to Maximize Well-Being and Boost Bottom Lines*. New York: McGraw Hill.
2. World Health Organization. (2019, May 28). *Burnout an 'Occupational Phenomenon': International Classification of Diseases*. https://www.who.int/news/item/28-05-2019-burn-out-an-occuapational-phenomenon-inernational-classification-of-diseases.
3. Gallup. (2021). *State of the Global Workplace 2021 Report*. Gallup Paper. https://gallup/com/2021/state-of-the-global-workplace-2021-report/.
4. Moss, J. (2025). *Unlocking Happiness at Work: How a Happiness Strategy Fuels Purpose, Passion, and Performance, 2nd ed*. London: Kogan Page.
5. Maslach, C, and Leither, MP. (2000). *The Truth About Burnout: How Organizations Case Personal Stress and What To Do About It*. San Francisco: Jossey-Bass.
6. Clifton, J, and Harter, J. (2021). *Wellbeing at Work: How to Build Resilient and Thriving Teams*. Washington: Gallup Press.
7. Edmondson, AC. (2018). *The Fearless Organization: Creating Psychological Safety in the Workplace for Learning, Innovation, and Growth*. Hoboken: Wiley.
8. Riess, H, and Neporent, L. (2018). *The Empathy Effect: Seven Neuroscience-Based Keys for Transforming the Way We Live, Love, Work, and Connect Across Differences*. Louisville: Sounds True.
9. Moss, J. (2021). *Beyond Burned Out*. HBR.org. https://www.hbr.org/2021/02/beyond-burned-out.
10. SAMHSA. (2016). *Creating a Healthier Life: A Step-By-Step Guide to Wellness*. https://library.samhsa.gov/product/creating-healthier-life-step-step-guide-wellness/sma16-4958.
11. Volk, JO, Schimmack, U, Strand, EB, Reinhard, A, Hahn, J, Andrews, J, Probyn-Smith, K, and Jones, R. (2024, May 10). Work-life balance is essential to reducing burnout, improving well-being. *Journal of the American Veterinary Medical Association*, 262(7), 950–957. https://doi.org/10.2460/javma.24.02.0135. PMID: 38729208.
12. Fisher, J, and Phillps, A. (2021). *Work Better Together: How to Cultivate Strong Relationships to Maximize Well-Being and Boost Bottom Lines*. New York: McGraw Hill.
13. Wellable. (2025). *2025 Employee Wellness Industry Trends Report*. https://www.wellable.co/resources/employee-wellness-industry-trends-reports/2025/.
14. Sapien Labs. (2024). *Work Culture & Mental Wellbeing*. https://sapienlabs.org/work-culture-mental-wellbeing-report-copy/.
15. World Economic Forum. (2025). *The Future of Jobs Report 2025*. https://www.weforum.org/publications/the-future-of-jobs-report-2025/.

16. Mental Health America. (2025). *Mind the Workplace 2024*. https://mhanational.org/2024-workplace-wellness-research.
17. Harvard Business Review. (2024). *Why Workplace Well-Being Programs Don't Achieve Better Outcomes*. https://hbr.org/2024/10/why-workplace-well-being-programs-dont-achieve-better-outcomes.
18. International Standard. (2021). *ISO 45003:2021: Occupational Health and Safety Management- Psychological Health and Safety at Work – Guidelines for Managing Psychosocial Risks, Edition 1*. https://www.iso.org/standard/64283.html.
19. Sovold, L., et al. (2021). Prioritizing the mental health and well-being of healthcare workers: An urgent global public health priority. *Front Public Health*, 9, 679397. https://doi.org/10.3389/fpubh.2021.679397.
20. Volk, JO, Schimmack, U, Strand, EB, Vasconcelos, J, and Siren, CW. (2020). Executive summary of the Merck animal health veterinarian wellbeing study II. *Journal of the American Veterinary Medical Association*, 256, 1237–1244. https://doi.org/10.2460/javma.256.11.1237.
21. Steffey, MA, Griffon, D, Scharf, VF, Risselada, M, Buote, NJ, and Zamprogno, H. (2023). Demographics of veterinary burnout and the effect and impacts of occupational stress on organizations such as education and practice: A narrative review. *Frontiers in Veterinary Science*, 10, 1184526. https://doi.org/10.3389/fvets.2023.1184526.
22. Steffey, MA, et al. (2023). A narrative review of the physiology and health effects of burnout associated with veterinarian-pertinent occupational stressors. *Frontiers in Veterinary Science*, 10, 1184525. https://doi.org/10.3389/fvets2023.1184525.
23. Neill, CL, Hansen, CR, and Salois, M. (2022). The economic cost of burnout in veterinary medicine. Frontiers in Veterinary Science, 9, 814014. https://doi.org/10.3389/fvets.2022.814104.
24. Miller, W, and Rollnick, S. (2012). *Motivational Interviewing: Helping People Change*, 3rd ed. New York City: Guilford Press.
25. Mental Health First Aid. https://www/mentalhealthfirstaid.org.
26. Personal communication, July 8, 2025.
27. Varble, D. (2022). *We are Treating Burnout All Wrong. Today's Veterinary Practice*. https://todaysveterinarypractice.com/personal-wellbeing/we-are-treating-burnout-all-wrong/.
28. Riess, H, and Neporent, L. (2018). The Empathy Effect: Seven Neuroscience-Based Keys for Transforming the Way We Live, Love, Work, and Connect Across Differences. Louisville: Sounds True.
29. Galaxy Vets. (2022). *A Call for Progress: Updating Veterinary Professional Oaths*. https://galaxyvets.com/media-kit/releases/updating-veterinary-professional-oaths/.
30. Hawley, Q. GetMotiVETed University's Veterinary Burnout P.R.E.P. Certification Program. https://www.getmotiveted.com.

6

Beyond the Summit

Support for the Trail of the Future

"Change begins at the end of your comfort zone."

– Roy T. Bennett

"The goal is to turn data into information, and information into insight."

– Carly Fiorina

During my training to become a certified neuro-informed coach, neuroscience findings and applications that impact human experience were explored in depth. With the advent of new technologies and the dedication of curious researchers, we increasingly have a much more accurate understanding of human neurobiological and energetic physiology and experience. Humans are gifted with many interconnected systems that keep us safe, provide multi-layered data that informs our thinking and behavior, and allows us to engage with energies inside and all around us. What is really cool is that we each have access throughout our lives to reflect, learn, and upskill our innate capacities to optimize our human experience. How?

- Learn more about your intuition systems and how to listen and more accurately interpret them. Intuition is 90% correct. It is our interpretations that are often wrong. Mind the inflow to improve the outflow.[1]
- Neuroplasticity is the ability to literally change our minds throughout the course of our lives. With intentional practices the

DOI: 10.1201/9781003475361-6

brain can grow and rewire to support healing, resilience, and fulfillment.
- Dr. Linda Page stated: "Neuroplasticity is a 6-syllable word for hope."[2] This quote is pointing to the "good news" that the brain can change and adapt through life, contributing to personal growth and transformation.
- Integration of our brain's systems offers optimization of our energy and amplifies our capacities to learn, connect, and innovate as individuals and in community.

Research supports your brain being capable of life-long learning and adaptation. Accessing and understanding the right data and tools for

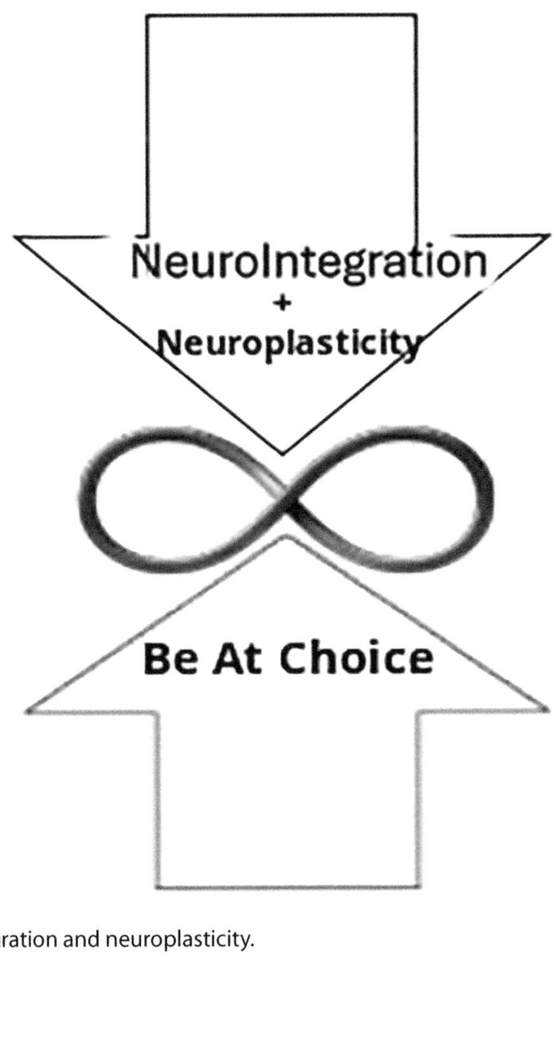

Figure 6.1 NeuroIntegration and neuroplasticity.

transformation when you are ready to "evolve" sets you up for impactful next steps on your unique journey of growth.

Two of my top values as a person and an emergency clinician are efficiency and humble curiosity. As a veterinary community, I would say we also share the strength of scientific scrutiny expediting critical thinking and discernment. Bringing these attributes together has resulted in this chapter of curated resources for you. Through my own research and experience, I have been exposed to a wide variety of assessments, behavior-changing tools/strategies, and trainings that have supported my personal and professional growth.

I only know what I know! So I reached out to many folks in our global veterinary family involved in culture change and veterinary wellness. All of us are dedicated to the adventure of learning and discovering as a supportive community as we courageously peek around the bend on this dynamic path of life. I am honored and deeply grateful to have met and been in conversation with so many of these individuals. Our wisdom is now being shared here with you to expedite your own hike up the learning trail and discerning what next summit calls you.

> "I can do things you cannot, you can do things I cannot; together we can do great things."
>
> – Mother Teresa: A celebration of unique strengths, this quote encourages embracing differences to achieve common goals.

> "Coming together is a beginning; keeping together is progress; working together is success."
>
> – Edward Everett Hale: It's a journey from the first step of coming together, through the trials of staying together, to the triumph of working as one.

Recommended Assessments and Psychometric Tools: Data Informs Change and Drives Innovation

We are fortunate to have an abundance of tools and training resources that can provide meaningful insights to contribute to individual and collective positive change and growth for veterinary professionals. Many of these are not exclusive to vet med but are very relevant to being human.

To make it easier for you to find what you need when you need it, I have "batched" the tools into generalized categories.

Who and How Am I? Self-Awareness to Unlock Potential

Personality Assessments

- DiSC Personality Assessment – https://www.discprofile.com/
- OCEAN model of personality – https://centerforpsychology.me/big-five-120/
- Myers-Briggs Type Indicator (MBTI) – https://themyersbrriggs.com

Evaluating and Understanding Strengths, Values, Biases

- Brene Brown's Dare to Lead – Living into our values – https://brenebrown.com/resources/dare-to-lead-list-of-values/
- Gallup – Clifton StrengthsFinder – rebranded to CliftonStrengths – https://www.gallup.com/cliftonstrengths/en/strengthsfinder.aspx
- Personal Values Assessment – https://personalvalu.es/
- Living into Your Values – https://brenebrown.com/resources/living-into-our-values/
- VIA Character Strengths Survey – https://www.viacharacter.org/
- Harvard Implicit Association Test – https://edib.harvard.edu/implicit-association-test-iat
- PQ "Saboteurs Assessment" – https://www.positiveintelligence.com/saboteurs/

Wellbeing Assessments

- Personal Sustainability Index (a test by Richard E. Boyatzis and Daniel Goleman) – https://www/keystepmedia.com – Boyatzis, R. E., Goleman, D., Dhar, U., and Osiri, J. K. (2021). Thrive and survive: Assessing personal sustainability. Consulting Psychology Journal: Practice and Research, 73(1), 27.
- AVMA Wellbeing Self-Assessment – https://myvetlife.avma.org/rising-professional/your-wellbeing/wellbeing-self-assessment
- AAVMC 2020 Wellbeing Assessment Tool – https://www.aavmc.org/resources/wellbeing/
- Self-Care Questionnaire (excerpted from "Resilience, Balance and Meaning" by Dr. Patricia Fisher) (PDF) https://www.stsconsortium.com/free-resources
- Perceived Stress Scale (PSS) – https://novopsych.com/assessments/well-being/perceived-stress-scale-pss-10/
- Warwick–Edinburgh Mental Wellbeing Scale (WEMWBS) – https://warwick.ac.uk/services/innovations/wemwbs
- Professional Quality of Life Survey for Healthcare Providers (ProQOL Health) – https://proqol.org/proqol-health-1 (*Note: This is*

a non-diagnostic, pulse point evaluation tool to gauge the current state of occupational wellness or dissatisfaction being experienced by a caregiver.)

- SAMHSA Wellness Wheel – 8 Dimensions of Wellness – https://www/samhsa.gov and www.samhsa.gov/wellness-initative – Supports reflection of how to take a more balanced approach to the 8 realms of wellness: physical, intellectual, emotional, social, spiritual, vocational, financial, and environmental.
- PERMA Model – https://positivepsychology.com/perma-model/
  - PermaProfiler – the_perma-profiler_101416.pdf
  - Workplace PERMA Profile – workplace_perma_profiler_102014.pdf
  - Purpose + PERMA Profiler – https://www.purposeplus.com/survey/perma-profiler
- The Flourishing Scale – https://positive.b-cdn/net/wp-content/uploads/The-Flourishing-Scale-Statements.png

Maslach Burnout Inventory

- Maslach Burnout Inventory (Free) – https://maslachburnout.com/
- Maslach Burnout Inventory (MBI) – Assessments, Tests | Mind Garden – https://www.mindgarden.com/117-maslach-burnout-inventory-mbi

Christina Maslach's Areas of Work-Life Survey

- Areas of Worklife Survey (AWS) – Assessments, Tests | Mind Garden – https://www.mindgarden.com/274-areas-of-worklife-survey
  - Early Identification of Burnout Risk: By pinpointing areas of dissatisfaction, organizations can address issues before they lead to full-blown burnout.
  - Improved Employee Wellbeing: Addressing the root causes of burnout helps create a more positive and supportive work environment.
  - Increased Retention Rates: When employees feel valued, empowered, and aligned with the organization's mission, they are more likely to stay.
  - Enhanced Productivity and Engagement: A healthier work environment leads to increased employee engagement and ultimately, higher productivity.

Burnout Assessment Tool (BAT)

- Burnout Assessment Tool (BAT) – https://burnoutassessmenttool.be/wp-content/uploads/2020/08/Test-Manual-BAT-English-version-2.0-1.pdf

Depression Anxiety and Stress Scale 21 (DASS-21)

- Perceived Stress Scale (PSS) – https://novopsych.com/assessments/well-being/perceived-stress-scale-pss-10/

Kessler Psychological Distress Scale (K10)

- Kessler Psychological Distress Scale – https://novopsych.com/assessments/outcome-monitoring/the-kessler-psychological-distress-scale-k10/ – A psychological screening tool designed to identify adults with significant levels of psychological distress. It is widely used in Australia and often used in primary care settings to identify people with clinically significant psychological distress.

Secondary Traumatic Stress

- Secondary Trauma Impacts Scale (STIS) – https://www.stsconsortium.com/_files/ugd/72c6ae_0b8492f6acb84a21b956f9bac457eb02.pdf
- Secondary Traumatic Stress Scale – https://www.stsconsortium.com/_files/ugd/72c6ae_96666786a8514dd9a9c8850c35e76598.pdf
- Secondary Traumatic Stress Scale for DSM 5 – https://www.stsconsortium.com/_files/ugd/72c6ae_d265add2633440caa2bc12828c190023.pdf

Resilience

Resilience scales are powerful tools for:

- Baseline diagnostics: Understanding the overall resilience levels of employees.
- ROI of wellbeing programs: Measuring the impact of employee wellbeing initiatives.
- "White spot analysis": Identifying areas where resilience training is needed.
- Training design: Creating tailored programs to build resilience across the organization.

Resilience isn't just a personal trait but a skill that can be developed and nurtured. Only by measuring it can interventions be created to enable individuals, teams, and organizations to sustainably thrive.

Resilience Scale for Adults (RSA)

- Resilience Scale for Adults (RSA) – https://www.envisionyourevolution.com/psychological-tests/styles-attitudes-tests/the-resilience-scale-for-adults/

The oldest scale and one that is still very much in use today is The Resilience Scale, developed by Wagnild and Young in 1993. The results have been found to positively correlate with physical health, morale, and life satisfaction, while negatively correlating with depression.

The scale is intended to measure resilience based on five essential characteristics:

1. Meaningful Life (or Purpose)
2. Perseverance
3. Self-Reliance
4. Equanimity
5. Existential Aloneness

Other Resilience Scales

- Brief Resilience Scale (BRS) – https://www.personalresilienceindicator.com/2025/04/24/brief-resilience-scale/
- Nicholson McBride Resilience Questionnaire (NMRQ) – https://blocksurvey.io/templates/resilience-scales/nicholson-mcbride-resilience-questionnaire
- Personal Resilience Indicator (PRI) – https://www.personalresilienceindicator.com/take-the-pri/ – Developed in 2021, PRI stands out as the most advanced resilience scale in 2025 per online reviews of these psychometric tools, merging decades of neuroscience with practical coaching applications. Developed by Mind Matters.

Who Are We and Where Are we Now on Our Path?

Culture Assessment and Change Management Support

- Fearless Organization – Psychological Safety for Self and Team – (https://fearlessorganizationscan.com)
- Measuring Psychological Safety – Psych Safety – (https://psychsafety.com/measure-psychological-safety/)
- Conflict Styles Assessment – https://www.usip.org/public-education-new/conflict-styles-assessment
- Conflict Management Styles Assessment – http://www.blake-group.com/sites/default/files/assessments/Conflict_Management_Styles_Assessment.pdf
- Thomas-Kilmann Conflict Mode Instrument (TKI) – Take the Thomas-Kilmann Instrument | Improve How You Resolve Conflict – (kilmanndiagnostics.com)

Professional Development: Self-Assessments to Inform Your "Next"

AVMA

Created by the Virginia-Maryland Regional College of Veterinary Medicine and the AVMA, these professional development resources are helpful in identifying what is most likely to satisfy you professionally.

- Self-Assessment Exercise – Begin by defining your career goals, strengths, limitations, and interests – https://www.avma.org/sites/default/files/resources/2014_06_VA-MD_Exercise.pdf
- Self-Assessment Summary – Mesh your career interests with your personality type to summarize what career field would be best for you – https://www.avma.org/sites/default/files/resources/2014_06_VA-MD_Exercise_Summary.pdf

Operators to Owners – Veterinary Medicine's Flourishing-Based Intellectual Community

- Self-assessment tools to discover your potential as a high performance veterinary medical practitioner – https://otovets.com/self-assessments/

Veterinary Hospital Managers Association's Empower U Self-Assessment

- Practice management professionals can now access a self-assessment tool that will allow them to evaluate their veterinary practice management knowledge. Empower U Self-Assessment test is a non-proctored (no time limits) examination modeled after the CVPM examination and is available to all practice management professionals (note: the tests are available 24/7 through the VHMA Webassessor, and the cost per test form is $69 USD per use) – https://www.vhma.org/resources/empower

VetSet2Go

- VetSet2Go – https://www.vetset2go.edu.au/modules/vetset2go/content/reflective-self-evaluation.pdf

Veterinary Management Group (VMG) Veterinary Leadership Standards Framework

- VMG Veterinary Leadership Standards Framework 2020 – https://vetmg/com/vmg-leadership-standards-framework-2020-download/
- Assess Your Own Leadership Skills Online Tool – https://form.jotform.com/202884505130348

Hogan Assessment

- Hogan Assessment – https://www.hoganassessments.com/talent-development/leadership-development/ – It's a personality instrument that has validity and reliability and is used quite often in leadership development to help leaders understand how others are experiencing them (Note: It does require a certified practitioner to debrief it).

Vark Learning Styles

- VARK Learning Styles – https://vark-learn.com – A multifactorial analysis utilized to match trainers and trainees based upon learning and instructing styles.

Organizational Health Assessment Tools

- Core Competencies in Trauma-Informed Supervision – This guide is intended to be a developmental STS assessment for supervisors, to help identify areas of need, and to guide the user to resources to strengthen those areas of competency.
- Secondary Traumatic Stress Core Competencies for Trauma-Informed Support and Supervision: Cross-Disciplinary Version | The National Child Traumatic Stress Network – https://www.nctsn.org/resources/secondary-traumatic-stress-core-competencies-for-trauma-informed-support-and-supervision-cross-disciplinary-version
- Low Impact Debriefing – Four steps to protect yourself from being "slimed" and to help minimize the risk of traumatizing your colleagues, friends, and family – https://www.stsconsortium.com/_files/ugd/72c6ae_44ccbd3eaff84301ba454b8cc9ed5214.pdf

PERMAH Wellbeing Survey and Professional Quality of Life Scale

- The "us" level refers to the culture of wellbeing in the workplace and the organization's responsibilities. It might consist of regularly assessing wellbeing to discover what is going well and where gaps in wellbeing and engagement exist. Validated tools, such as the PERMAH Wellbeing Survey (https://permasurvey.com) and the Professional Quality of Life Scale (https://proqol.org/proqol-health-1) assist leaders by giving actionable data on wellbeing and burnout.

Certifications and Training Programs

- **MentorVet Veterinary Mentor Certificate Program** – https://www/mentorvet.net – A yearlong robust evidence-based veterinary mentor certificate program open to any individual associated with the veterinary profession interested in:
  - Becoming a more effective mentor
  - Learning skills to create and evaluate intentional mentorship structures
  - Connecting with other mentors
- **University of Tennessee – Veterinary Social Work** – https://vetsocialwork.tennessee.edu/ – The Center for Veterinary Social Work offers training and education in the interprofessional space of veterinary social work for anyone with or pursuing a health-related license. Individuals who have been in animal-related professions for over 5 years can also apply for a waiver of the health-related license requirement on a case-by-case basis. If you are not a social worker, you may even decide to pursue a master of science in social work as part of your veterinary social work journey!
  The 4 Areas of Veterinary Social Work Modules
  - Intentional WellBeing
  - Animal-Related Grief and Bereavement
  - Animal-Assisted Interventions
  - The Link Between Human and Animal Violence
- **University of Tennessee – Veterinary Human Support Certificate Program** – https://vetsocialwork.tennessee.edu/veterinary-human-support-certificate/ – The Veterinary Human Support Certificate provides education for animal related professionals on the four areas of Veterinary Social Work. This certificate program is for animal-related professionals who have some form of health related state registration or license (DVM, LVT, etc.).
- **GetMotiVETed!** – https://getmotiveted.com/burnout-certification

In the words of co-founder, Dr. Quincy Hawley:

"Veterinary medicine is a calling built on compassion, but compassion without preparation can leave professionals vulnerable to burnout. That's why we created the **Veterinary Burnout P.R.E.P. Certification Program** through Get MotiVETed University, where the veterinary professionals can earn their Certificate of Veterinary

Burnout Preparedness (CVBP). Rather than waiting for crisis, we equip veterinary professionals with evidence-based and battle-tested skills grounded in four pillars: Prevention, Resilience, Empowerment, and Preparedness.

What makes the program stand out are the personal transformations we've witnessed and the measurable results. Using the Professional Quality of Life (ProQOL) assessment, we compared metrics before and after certification. The outcomes we are promising:

- High burnout dropped from 25.4% to 4.3%.
- High compassion satisfaction nearly doubled, from 33% to 64.5%.
- High secondary traumatic stress decreased from 19% to 4.3%.
- Perceptions of strong support more than doubled.

Every metric trended toward improved wellbeing. These results affirm that when veterinary professionals prepare intentionally, they protect their own wellbeing and strengthen the culture of their teams. Our motto is, "Veterinary burnout is REAL, but it doesn't have to be your REALity!" It's our promise to help make true wellbeing the reality for as many veterinary professionals as possible!"

- **Mental Health First Aid** – https://www.mentalhealthfirstaid.org – International program to teach skills to respond to the signs of mental illness and substance abuse.

Stress First Aid (SFA) is a framework for peer support and self-care for health care professionals. The SFA framework is parallel to how a clinician or first-responder approaches physical first aid – intervening when needed to remove the stressor, prevent further harm, and promote recovery for those who have been strongly affected by significant stress.[1] Individuals may employ this framework when supporting peers or if they are aware of a potential injury in themselves, whereas organizations may train a team of health care professionals to administer SFA to their colleagues.

Stress First Aid (SFA) for Healthcare Workers was developed by Drs. Patricia Watson and Richard Westphal as an adaptation to the original program designed for military populations.

- **Stress First Aid for Veterinary Professionals with Melyssa Allen,** MA, NBC-HWC, DIPACLM – https://veterinarywellbeingbuddy.com/stress – While a specific veterinary protocol has not yet been developed using the SFA program, the teachings can be adapted to veterinary team environments. Using the stress continuum, veterinary professionals can check on their stress zone and also ask their coworkers and teams about which stress zone they are experiencing.

- **ASK Suicide-Screening Questions (ASQ) Toolkit** – https://www.nimh.nih.gov/research/research-conducted-at-nimh/asq-toolkit-materials – (Authoring Agency: National Institute of Mental Health – NIMH) – The Ask Suicide-Screening Questions tool is a brief (20-second) assessment that health care professionals can administer in a variety of settings (emergency department, inpatient medical unit, primary care clinics) to gauge suicide risk in patients. The toolkit website explains how to administer and respond to screening test results.
- **QPR Gatekeeper Training** – https://qprinstitute.com/individual-training – QPR training is a suicide prevention program that teaches individuals how to recognize warning signs, question those at risk, persuade them to seek help, and refer them to appropriate resources. QPR stands for Question, Persuade, Refer. It is an evidence-based training program designed to help individuals recognize the signs of a suicide crisis and take action to prevent it.

AVMA

- **Train the Trainer for Wellbeing at Work Workplace Wellbeing Certificate Program** – https://axon.avma.org/local/catalog/view/product.php?productid=22 – The program starts with the module "Creating a Culture of Wellbeing" with Jen Brandt, PhD, AVMA director of wellbeing and diversity. Then they finish up the following in any order they want:
  - "How to Request, Receive and Give Feedback" with Dr. Brandt
  - "Transforming Conflict" with Dr. Brandt and Elizabeth Strand, PhD, founding director of veterinary social work at the University of Tennessee College of Veterinary Medicine
  - "QPR Assessment" (risk assessment and management of suicide risk) with the QPR Institute
  - "Diversity and Inclusion" with Lisa Greenhill, MPA, EdD, AAVMC senior director for research and diversity, and Dane Whitaker, DVM, president-elect for Pride Veterinary Medical Community (formerly the Lesbian and Gay Veterinary Medical Association).
  - **Purdue Global Graduate Certificate in Diversity, Equity, and Inclusivity** – https://www.purdueglobal.edu/degree-programs/professional-studies/diversity-equity-inclusivity-dei-graduate-certificate/

CVMA

- **Canadian Veterinary Medical Association – The Working Mind** – https://www.canadianveterinarians.net/veterinary-resources/veterinary-health-and-wellness-resources/the-working-mind/ – The Canadian Veterinary Medical Association (CVMA) has recently partnered with Health insurance provider Petsecure Pet Health Insurance to create The Working Mind, a novel mental health program for veterinary teams.

The Working Mind is structured as a training course designed to address the inherent workplace stress faced by individuals in the veterinary field. Program attendees, including both veterinary managers and employees, have reported improved attitudes toward mental health distress, increased resiliency during difficult times, and an overall increase in mental and emotional wellbeing.

- **TEND Academy** – https://www.tendacademy.ca – TEND offers free resources and training to address the complex needs of high stress, trauma-exposed workplaces.
  - Trauma-Informed Practice
  - Secondary Traumatic Stress
  - Empathic Strain for Caregiving Professionals
  - Debriefing and Mutual Support

Other Certifications

- The Centre for Healing's Trauma Informed Certification – https://www.thecentreforhealing.com – "When You Understand Trauma & How To Effectively Treat It, You Can Change The World. We Help Coaches, Therapists, Healers & Everyday People Turn Trauma into Triumph through our Certified Practitioner Trainings & Educational Courses"
- RECOVER CPR Certification – https://recoverinitiative.org
- FearFree Certifications – https://fearfree.org
- NAVC Certifications – https://navc.com – Includes credentialed programs in leadership, business, and technician education.
- BlendVet Achiever Program – https://www.blend.vet/licensed-content – The blendVET Licensed Partner Program offers veterinary organizations, large and small, access to high-quality, ready-to-use DEIB (Diversity, Equity, Inclusion, and Belonging) training designed specifically for veterinary teams. Through this program, partners can seamlessly integrate our content into their own learning platforms,

empowering their teams with flexible, impactful education on topics like unconscious bias, microaggressions, and inclusive leadership.
- AVTE Fellowship Program (FVTE) – https://www.avte.net/fellows – A transformative initiative aimed at recognizing and honoring the dedication and expertise of veterinary technician educators in the United States and Canada. This new designation bolsters educators' credibility and influence within all corners of veterinary education, elevating their voice and stature to their rightful place.

Blue EQ

As shared by Master Facilitator for BlueEQ, Kelly Newman:

> BlueEQ is a global leader in Emotional Intelligence (EQ) and Psychological Safety training, providing research-based tools and transformative learning experiences that empower individuals, strengthen teams, and build thriving organizational cultures. With roots in both science and practice, BlueEQ equips leaders with the skills to foster environments where people feel safe to contribute, challenge, and innovate – core elements of psychological safety that drive both well-being and performance.
>
> Our work spans industries, cultures, and professions, including health care and veterinary medicine, where professionals face unique pressures of compassion fatigue, client demands, and team dynamics. By helping individuals understand and expand their emotional intelligence, BlueEQ enables veterinarians, technicians, managers, and leaders to respond to stress with resilience, communicate more effectively, and build trust across every level of their organizations.
>
> What BlueEQ Offers:
>
> - Individual Development: Our validated Emotional Intelligence Self-Assessment provides personalized insights into strengths and blind spots, guiding participants toward practical action steps that improve empathy, adaptability, and stress management.
> - Team Growth: Through interactive workshops, teams learn how to move beyond conflict avoidance and toward constructive dialogue, resulting in stronger collaboration and problem-solving.

- Leadership Training: BlueEQ's leadership development programs focus on building high-trust relationships, balancing accountability with compassion, and modeling the psychological safety that inspires teams to do their best work.
- Culture Creation: Organizations partner with BlueEQ to intentionally design cultures that are inclusive, innovative, and resilient. By embedding psychological safety and EQ into the DNA of daily operations, leaders create workplaces where people thrive.

Veterinary professionals around the world are discovering that technical skills alone are not enough to sustain long, healthy, and fulfilling careers. Success and wellbeing increasingly depend on the ability to manage emotions, build relationships, and lead with influence. BlueEQ exists to make that possible.

To learn more about assessments, workshops, and leadership solutions, visit www.BlueEQ.com.

Lincoln Institute of Veterinary Business
Lincoln Institute of Veterinary Business is a leading provider of professional and leadership development for veterinary practitioners and business owners across Australia, New Zealand and globally. Per Dr. Mick Powell, BVSc (Hons), Director:

> We recognise that veterinary medicine is a calling – one that draws compassionate, intelligent, and dedicated people who want to make a difference for animals, their owners, and their communities. We understand that while work is deeply meaningful, it is also uniquely demanding. Long hours, emotional stress, financial pressures, and ever-rising client expectations can leave even the most committed practice owners, veterinarians, and support teams feeling overwhelmed, undervalued, or burnt out. Too often, individuals with extraordinary skills and passion leave the profession prematurely, not because of the medicine, but because of the toll of the work environment.
>
> At Lincoln, we believe it doesn't have to be this way. Founded with a vision to create a brighter future for the veterinary profession, we provide veterinary practitioners with the non-technical skills – leadership, communication, resilience, and emotional intelligence – that turn challenges into opportunities and workplaces into communities where people flourish.

What sets Lincoln apart is our recognition that veterinary success goes far beyond clinical expertise. Great medicine requires great teams, and great teams are built on trust, clarity, and strong leadership. By combining proven business principles with a deep understanding of the realities of veterinary life, we empower practice owners, managers, veterinarians, veterinary nurses, technicians, and front of house team members to build workplaces that are sustainable, successful and deeply rewarding.

We deliver this through evidence-based programs, coaching, masterclasses, webinars, and live events – grounded in lived experience and supported by pioneering global research – that create genuine change, not just in skills and strategies, but in mindset and culture. Our approach is practical, data-driven, and tailored to the veterinary context, ensuring everything learned can be applied to everyday practice.

Our impact is profound. The practices we work with consistently see improved team harmony, stronger financial performance, and higher levels of client satisfaction. Individuals report greater confidence, resilience, and satisfaction in their work, leading to longer, healthier, and more fulfilling careers and strengthening retention. In short, they rediscover why they chose this profession in the first place with renewed purpose.

Importantly, in a landmark development for the profession, all Lincoln programs are now formally recognised with a university-level credential from Southern Cross University. This strategic partnership ensures that veterinary professionals undertaking Lincoln training are awarded a jointly issued certificate that carries the credibility, quality assurance, and academic rigour of one of Australia's leading public universities.

At its core, Lincoln Institute of Veterinary Business is not just an education provider – it is a movement to reimagine what it means to truly thrive in veterinary practice.

To learn more, visit our website: www.lincolninstitute.com.au

Leadership Development

- Jon Gordon's Power of Positive Leadership Training (more on this later) – https://jongordon.com
- Center for the Advancement of WellBeing at George Mason University – https://wellbeing.gmu.edu/certificate-program
- Performance Management Consultants Workshop/Course – https://www/pmctraining.com/training/team-training

- Co-Active Training Institute (coaching and leadership development) – https://coactive.com
- BeAbove Leadership (advanced coach training) – https://www.beaboveleadership.com

Professional Development and Human Support Resources

Support clarity, growth, team harmony, and sustainable professional satisfaction.

Learning Resources and "Classrooms"

- Brene Brown's Dare to Lead Hub – https://www.brenebrown.com/hubs/dare-to-lead/
- Amy C Edmondson's Fearless Organization – https://fearlessorganizationscan.com
- Substance Abuse and Mental Health Services Administration (SAMSA) and other human medicine professional resources for caregiver support – (https://www.samhsa/gov)
- Polyvagal Wisdom – Deb Dana, Stephen Porges, Andrew Huberman
- Neuroplasticity and the Growth Mindset – Adam Grant, Rick Hansen, Carol Dweck

Mindfulness Practices/Resources

Far too many valuable and practical resources and hubs of wisdom to list here, but to get started, check out:

- Mindful Leader's Meditate Together – https://www.mindfulleader.org/meditate-together
- InStill Mindfulness online community – https://www.community/instillmiindfulness.org
- Global Online Meditation Community with U Mass Memorial Health – https://www.ummhealth.org
- Together Mindful – https://ww/togethermindful.com
- Greater Good – https://www/greatergood/berkeley.edu
- mindful – https://www.mindful.org

Mindfulness Based Stress Reduction (MBSR)

- Mind Matters Initiative's mindfulness-based stress management program – www.thewebinarvet.com
- Mindful Leader Institute – https://www/mindfulleader.org

- VetLife – https://vetlife.org.uk/sdm_downloads/burnout-moral-injury-and-compassion-fatigue-resource/ – Support for Veterinary Community: Burnout, moral injury, and compassion fatigue resource (2025)
- The PERMA + 4 Framework: A New Approach to WellBeing at Work | Psychology Today Source: Donaldson, Van Zyl, and Donaldson (2022) – https://psychologytoday.com/us/blog/happybytes/202307/the-perma4-framework-a-new-approach-to-well-being-at-work

Don't Stop Now!

Stay Curious and Motivated for the Last Leg of this Journey Together

This is not by any means an exhaustive list, and there may be resources available that I simply did not know of or that were not on our collective radar. Also, there are so many wonderful surveys and assessments that veterinary colleagues have created as part of their individual professional work. Those

Figure 6.2 Continuing the veterinary wellness trek together.

resources would be found within the training that they provide and/or on their respective websites.

With that said, let's continue to build our community and collective strength on the path toward our current professional "summit" and beyond in Chapter 7!

It is my pleasure to share with you a wonderful sampling of amazing folks who are change agents in the veterinary profession and are our healthy culture guides. What an uplifting and inspirational opportunity to share with you as well a nourishing platter of yummy resources to feed your hearts and minds as we continue the veterinary wellness trek together as a global community.

References

1. Kotler, S, Parvizi-Wayne, D, Mannino, M, Friston, K. (2025). Flow and intuition: A systems neuroscience comparison. *Neurosci Conscious*. Vol 2025:1, niae040. https://doi.org/10.1093/nc/niae040.
2. Page, L. Personal communication, June 10, 2025.

7

Uplifting Community and Inspirational Connections

The Fuel to Keep Us Moving Forward

Welcome to the Summit! This chapter serves as a curated guide to the many supports, initiatives, and learning opportunities available to veterinary professionals to continue to guide reflection, growth, and future path planning individually and collectively.

The resources span mental health services, organizational wellness initiatives, professional development programs, DEI-focused organizations, training tools, podcasts, books, and more. Each entry includes the name of the resource, its website, and where available, a brief description contributed by colleagues or sourced from the organization.

Please note that a PDF is available online, downloadable from the book's webpage at www.routledge.com/9781032753928, which contains many additional resources and links. This chapter offers a curated selection. While the list is not exhaustive, it is designed to help readers quickly locate and explore supportive spaces, trusted references, and communities that resonate with their needs.

You are encouraged to mark resources of personal relevance and return to this section as your career and wellness journey continues. There is a saying from East Africa that to move quickly, move alone and to move far, move together. Choose your path and your pace. You've got this!

This collection of resources and voices represents a living toolkit for veterinary professionals. As the profession evolves, so too will the supports, innovations, and

DOI: 10.1201/9781003475361-7

communities available. Readers are encouraged to revisit this section and the online PDF often, to share what resonates, and to continue shaping a veterinary culture where wellbeing and sustainability are collective priorities.

Note: URLs are included in full to ensure accessibility in both print and e-book formats.

Section 1: Mental Health Support (Crisis & Immediate Assistance)

988 Suicide & Crisis Lifeline (U.S.)

Provides 24/7 free and confidential emotional support for people experiencing suicidal thoughts, mental health crises, or emotional distress. Offers phone, text, and chat access to trained crisis counselors.

Website: https://988lifeline.org

Crisis Text Line (U.S., UK, Canada, Ireland)

Offers free, 24/7 confidential support from trained crisis counselors via text message for individuals experiencing acute stress, panic, or emotional overwhelm. Designed for rapid, anonymous assistance during difficult moments.

Website: https://www.crisistextline.org

Employee Assistance Programs (EAPs)

Confidential workplace-based services that provide short-term counseling, crisis intervention, and referrals for personal, emotional, and workplace challenges. Often include free initial sessions for employees and family members.

Access: Contact your employer's EAP provider.

National Alliance on Mental Illness (NAMI) Helpline (U.S.)

Provides free support, information, and guidance for individuals and families facing mental health concerns. Offers a national help center and local chapter programs.

Website: https://www.nami.org/help

Not One More Vet (NOMV) – Lifeboat Support

Provides anonymous online peer support for veterinary professionals experiencing crisis, emotional overwhelm, or burnout. Moderated by trained volunteers and designed to offer connection and safety during vulnerable moments.

Website: https://www.nomv.org

Direct Support Page: https://www.nomv.org/resources/lifeboat/

SAMHSA National Helpline (U.S.)

A free, confidential 24/7 referral and information service for individuals facing substance-use disorders, mental health challenges, or co-occurring conditions. Offers multilingual assistance.

Website: https://www.samhsa.gov/find-help/national-helpline

Vets in Mind – Mental Health Triage (International)

Provides free digital triage tools for veterinary professionals experiencing mental health concerns, helping individuals identify risks and connect to appropriate resources.

Website: https://vets-in-mind.org/triage-tools/

Veterinary Mental Health Initiative (VMHI) – Shanti Project (U.S.)

Offers no-cost, confidential therapy by licensed clinicians trained to work with veterinary professionals. Services include short-term therapy and facilitated groups.

Website: https://www.shanti.org/programs-services/veterinary-mental-health-initiative/

Section 2: Organizational Wellness Resources

American Association of Veterinary Medical Colleges (AAVMC)

Provides wellbeing initiatives, leadership development programs, and institutional resources that support thriving academic environments for students and faculty.

Website: https://www.aavmc.org/programs/wellbeing/

American Veterinary Medical Association (AVMA)

Offers comprehensive wellbeing tools, organizational culture resources, and research-based support services for veterinarians and veterinary teams.

Website: https://www.avma.org/resources-tools/wellbeing

Australian Veterinary Association (AVA) – VetHealth

Provides member-exclusive mental health support, crisis resources, and wellbeing education for veterinary professionals across Australia.

Website: https://www.ava.com.au/member-services/vethealth/

Canadian Veterinary Medical Association (CVMA)

Curates mental health and wellbeing resources, including webinars, tools, and national support services for Canadian veterinary professionals.

Website: https://www.canadianveterinarians.net/veterinary-resources/veterinary-health-and-wellness-resources/

National Association of Veterinary Technicians in America (NAVTA)

Provides wellbeing resources, professional development opportunities, and tools that support the growth and wellness of veterinary technicians.

Website: https://www.navta.net/page/wellbeing

Veterinary Nurses Council of Australia (VNCA)

Offers professional development, wellbeing education, and support for veterinary nurses across Australia.

Website: https://www.vnca.asn.au/wellness-hub/education/

World Small Animal Veterinary Association (WSAVA)

A global organization offering mental health initiatives, continuing education, and wellbeing-focused projects supporting veterinarians worldwide.

Website: https://wsava.org

Section 3: Virtual Mental Health Support

7 Cups

An online platform offering free emotional support from trained volunteer listeners, plus optional low-cost therapy with licensed clinicians. Designed for anonymity and accessibility.

Website: https://www.7cups.com

BetterHelp

Provides global virtual counseling with licensed therapists via messaging, phone, and video. Offers flexible scheduling and a broad range of mental health specialties.

Website: https://www.betterhelp.com

E-Counseling.com

Provides reviews, comparisons, and background information on virtual therapy platforms, allowing users to make informed choices about online mental health care.

Website: https://www.e-counseling.com

Kite Veterinary (Kite for Vets)

A digital wellbeing platform tailored to veterinary professionals, offering resilience-building tools, structured wellbeing programs, and personal development resources.

Website: https://www.thekiteprogram.com/kite-industry/kite-for-vets/

Lyra Health

A workforce mental health provider offering rapid access to virtual therapy, coaching, and mental health navigation tools for employees.

Website: https://www.lyrahealth.com

Talkspace

Provides virtual therapy and psychiatry services through text, audio, and video messaging, offering flexible, remote mental health support.

Website: https://www.talkspace.com

TogetherAll

A clinically moderated online community offering peer support, self-guided courses, and wellbeing tools in an anonymous and inclusive virtual environment.

Website: https://togetherall.com

Section 4: Additional Veterinary Wellbeing Resources

AAHA Culture Initiative

Provides tools, training modules, and frameworks to help veterinary practices build psychologically safe, mission-driven, collaborative workplace cultures.

Website: https://www.aaha.org/practice-resources/healthy-workplace-culture

I Matter

A Canadian program supporting mental health literacy and wellbeing among veterinary professionals, offering free tools and educational materials.

Website: https://www.i-matter.ca

Mighty Vet

A nonprofit offering mentorship, wellbeing education, leadership development, and suicide-prevention resources for veterinary professionals.

Website: https://www.mightyvet.org

Mind Matters Initiative (UK)

A Royal College of Veterinary Surgeons (RCVS) program promoting mental health awareness, training, and support across the UK veterinary sector.

Website: https://www.vetmindmatters.org

Veterinary Mental Health Initiative (VMHI)

Provides no-cost confidential therapy and support groups for veterinary professionals, emphasizing trauma-informed care and emotional wellbeing.

Website: https://www.shanti.org/programs-services/veterinary-mental-health-initiative/

Section 5: Websites With Wellbeing Information and Free Resources

AVMA Wellbeing Resources

Offers wellbeing, resilience, communication, financial, and workplace culture tools to support veterinary professionals.

Website: https://www.avma.org/resources-tools/wellbeing

Barrington Centre

Provides wellbeing, communication, and psychological safety resources supporting emotionally healthy workplaces.

Website: https://www.barringtoncentre.com/

CDC Worksite Health ScoreCard

A comprehensive organizational tool for assessing and improving workplace health initiatives, including mental wellbeing.

Website: https://www.cdc.gov/workplacehealthpromotion/initiatives/healthscorecard/index.html

Compassion Course for Veterinarians

An evidence-informed training resource supporting compassion skills, emotional regulation, and resilience in veterinary care.

Website: https://openresearch.surrey.ac.uk/esploro/outputs/coursemodule/99885965702346

Mental Health Foundation – Te Whare Tapa Whā

A holistic wellbeing model integrating mental, physical, spiritual, and community wellness practices.

Website: https://mentalhealth.org.nz/te-whare-tapa-wha

Mindfulness Resources:

- **Greater Good Science Center** – https://www.greatergood.berkeley.edu/topic/mindfulness
- **MBSR** – https://mbsrtraining.com
- **PositivePsychology.com** – https://www.positivepsychology.com/mindfulness
- **UCLA Mindful** – https://www.uclahealth.org/uclamindful/guided-meditations

NOMV CLEAR Blueprint

Provides organizational frameworks for building psychologically safe, sustainable veterinary workplaces.

Website: https://nomv.org/clear-blueprint/

Psychological Safety Resources (Iterum)

A collection of articles, tools, and training resources on fostering psychological safety at work.

Website: https://psychsafety.com/search/

Self-Compassion Resources:

- **Center for MSC** – https://centerformsc.org
- **Self-Compassion.org** – https://self-compassion.org
- **FECAVA Wellbeing Resources**:
  https://www.fecava.org/clinicians/mental-health-and-well-being/
  https://www.fecava.org/clinicians/mental-health-and-well-being/mh-wb-hub/

Veterinary Visionaries

A collaboration of veterinary organizations producing collective-action well-being initiatives and open-source tools.

Website: https://veterinaryvisionaries.org

Vet Life UK Resources

Provides financial, mental health, and peer-support services for UK veterinary professionals.

Website: https://www.vetlife.org.uk

Section 6: Trauma-Informed Care

Center for Nonviolent Communication

Provides resources, training, and frameworks for compassionate communication and conflict de-escalation.

Website: https://www.cnvc.org

The Centre for Healing

Offers courses and training on trauma recovery, emotional healing, and somatic therapies.

Website: https://www.thecentreforhealing.com/

NOMV Trauma Webinar Series

A three-part veterinary-focused webinar series introducing trauma concepts, responses, prevention, and healing:

- Part 1: https://youtu.be/oup2_he07kc
- Part 2: https://youtu.be/ID0UkBLz7Yg
- Part 3: https://youtu.be/3yuRy85YyZ4

SAMHSA Trauma-Informed Care Resources

Provides national guidelines, frameworks, and best practices on trauma-informed approaches across care settings.

Website: https://www.samhsa.gov

Secondary Traumatic Stress Consortium

Shares research, best practices, and tools for managing STS in caregiving professions.

Website: https://www.stsconsortium.com/

TEND Academy

Offers training and resources on burnout, secondary traumatic stress, and high-stress work environments.

Website: https://www.tendacademy.ca

Section 7: Community Support Spaces

Decade One Vet

Provides mentorship, community, and support for early-career veterinarians navigating the transition into practice.

Website: https://www.decadeonevet.com/

DVMoms
A supportive community dedicated to veterinary mothers, offering connection, shared experiences, and resources.
Website: https://www.thedvmoms.com

Relief Veterinary Medical Association
Supports relief veterinarians through community-building, professional development, and advocacy.
Website: https://www.reliefvma.org

Sustainability in Equine Practice Seminar & Retreat
Provides education, community support, and wellbeing tools for equine practitioners seeking sustainable careers.
Website: https://www.sepseminars.com/

Vet Mums
A peer-support community for veterinary mothers offering connection, discussion, and shared lived experience.
Website: https://www.facebook.com/groups/794407730649161

Section 8: Professional Development Opportunities and Resources

Centre for Veterinary Education (CVE)
Provides internationally recognized continuing education for veterinary professionals through online courses, workshops, and practical intensives. CVE focuses on evidence-based skills development to support clinical excellence and career-long growth.
Website: https://www.cve.edu.au
Hound
A veterinary professional development platform offering leadership training, communication courses, and career advancement resources. Designed to support long-term, sustainable career pathways in veterinary medicine.
Website: https://www.hound.vet

Vet Education
Offers global online continuing education webinars, short courses, and learning modules for veterinary professionals. Provides accessible, high-quality clinical and professional development content.
Website: https://www.veteducation.com

VMAE – Power of 10 Program

A national leadership development initiative designed for early-career veterinarians, focusing on communication, self-awareness, and leadership competencies. Hosted through state VMAs.

Website: https://vmae.org/resources

VetGirl

A digital CE platform providing on-demand lectures, podcasts, and clinical training for busy veterinary professionals seeking efficient, high-quality learning.

Website: https://vetgirlontherun.com

Section 9: Financial Support

AVMA Personal Finance Resources

Provides tools, calculators, and guidance to help veterinarians manage debt, budgeting, insurance, and long-term financial planning.

Website: https://www.avma.org/resources-tools/personal-finance

Florida Veterinary Advisors

A financial planning firm specializing in veterinary careers, offering personalized financial strategies, student loan guidance, and wealth-building education.

Website: https://flveterinaryadvisors.com

Richer Life DVM – Grace Kim

Provides resources that help veterinarians build financial confidence through education, values-based budgeting, and long-term financial wellness.

Website: https://richerlifedvm.com

Sabadoodle – Mike Sabatino

Offers accessible financial education and tools that help veterinary professionals manage money with confidence, reduce stress, and build sustainable financial habits.

Website: https://www.sabadoodle.com

Veterinary Debt Initiative

A collaborative initiative to address student debt in veterinary medicine through education, advocacy, and financial solutions.

Website: http://www.vmae.org/veterinary-debt-initiative

VIN Student Debt Center

Provides calculators, educational modules, and guidance for managing and understanding student loan obligations across career stages.

Website: https://www.vin.com/studentdebtcenter

Section 10: Human Resources & Practice Management

2 Manage Vets Consulting – Debbie Boone

Offers consulting services in customer service, workflow efficiency, and team development to support healthier, high-performing veterinary practices.

Website: https://www.debbiebooneconsulting.com

Digital Empathy

Provides website development and communication tools designed to improve client understanding, emotional connection, and user experience.

Website: https://www.digitalempathy.com

Eric Garcia – IT & Digital Strategy Consulting

An internationally recognized consultant helping veterinary teams improve client communication, cybersecurity, and digital workflow efficiency.

Website: https://ericgarciafl.com

Happy Doc

A documentation platform helping veterinarians streamline recordkeeping, reduce administrative burden, and reclaim time for clinical work.

Website: https://www.happydoc.ai

Kindred.Vet – Dr. Christie Cornelius

Supports end-of-life service providers with training, consulting, and compassionate practice development resources.

Website: https://www.kindred.vet

MKB Strategic Consulting – Melanie Barham

Provides operations, project management, and training solutions for veterinary organizations, supporting effective systems and sustainable workflows.

Website: https://www.mkbstrategic.com

National Veterinary Solutions

Offers HR support, staffing solutions, and operational resources for veterinary practices seeking organizational stability and growth.

Website: https://www.nationalveterinarysolutions.com

ScribeNote

An AI-powered veterinary scribing tool that helps streamline patient notes, improve documentation accuracy, and reduce clinician burnout.

Website: https://www.scribenote.com

Talkatoo

A speech-to-text dictation solution designed specifically for veterinary professionals to increase efficiency and reduce documentation fatigue.

Website: https://www.talkatoo.com

VHMA – Veterinary Hospital Managers Association

A national association offering CE, certification, HR training, and leadership development for practice managers.

Website: https://www.vhma.org/home

Zach Mercurio, PhD

An international expert in organizational purpose, belonging, and meaningful work who provides training and resources for building purpose-driven teams.

Website: https://www.zachmercurio.com

Section 11: Veterinary Practice / Team Culture Consultants & Training

Empowering Veterinary Teams (EVT)

Provides training, consulting, and professional development workshops focused on team communication, leadership, and psychologically safe work environments.

Website: https://www.empoweringveterinaryteams.com

Flourish Veterinary Consulting

A consulting group specializing in positive psychology, workplace wellbeing, and evidence-based culture transformation within veterinary practices.

Website: https://flourish.vet

Full Circle Lab – Crystal Stokes

Offers leadership coaching, organizational systems support, and human-centered transformation work grounded in neuroscience and emotional intelligence.

Website: https://www.fullcirclelab.com

Human-Savvy – Liv Oginska

Provides emotional intelligence training, conflict management workshops, and leadership development tailored to veterinary teams.

Website: https://behumansavvy.com

MKB Strategic Consulting

Delivers systems optimization, large-scale project management, facilitation, and organizational design support for veterinary teams.

Website: https://www.mkbstrategic.com

National Veterinary Solutions

Provides staffing support, team training, and operational solutions to improve workplace culture and team cohesion.

Website: https://www.nationalveterinarysolutions.com

Vet Empowered

A coaching and mentoring organization helping veterinarians build clarity, confidence, and empowered decision-making across all career stages.

Website: https://www.vetempowered.com

Vet Partners

A professional association of consultants, advisors, and experts supporting veterinary business success through education and best practices.

Website: https://vetpartners.org

Veterinary Team Training (VTT)

Provides training, consulting, and educational content to help teams improve communication, workflow, and collaboration.

Website: https://veterinaryteamtraining.com

VIBE – Veterinary Integrated Burnout Elimination

A program rooted in neuroscience and behavior change, helping veterinary professionals and teams prevent and recover from burnout.

Website: https://www.veterinaryburnout.com.au

Section 12: Veterinary Mentoring Spaces

blendVet

A mentorship and leadership development program advancing diversity, equity, and inclusion in veterinary medicine through training, workshops, and community building.

Website: https://www.blend.vet/

Lincoln Institute

Provides leadership training, team development programs, and veterinary-specific coaching designed to build emotionally intelligent, effective leaders.

Website: https://lincoln-institute.com

MentorVet

A national mentorship and professional support program for early-career veterinarians, offering training, coaching, and wellbeing-focused resources.

Website: https://www.mentorvet.net

Pawsibilities Vet Med

Supports diversity in the veterinary profession by providing mentorship, resources, and community for students and professionals from underrepresented groups.

Website: https://www.pawsibilitiesvetmed.com

Ready, Vet, Go.

A coaching and mentorship program providing early-career veterinarians with essential non-clinical skills to support confidence and professional wellbeing.

Website: https://www.readyvetgo.co

Section 13: Culture Change Agents

Affinity Futures

Provides foresight, strategic design, and culture transformation consulting to help veterinary organizations build resilient, future-ready systems.

Website: https://affinityfutures.com/

blendVet

A DEI-focused program promoting inclusive leadership, cultural awareness, and equitable practices across veterinary teams.

Website: https://www.blend.vet/

Diversify Veterinary Medicine Coalition (DVMC)

A coalition advancing equity, inclusion, and belonging by connecting organizations committed to diversifying the veterinary profession.

Website: https://diversifyvetmed.org

Flynn's Walk

An Australian charity promoting mental health awareness, suicide prevention, and community connection within veterinary medicine.

Website: https://www.flynnswalk.com.au/finding-support

GCBU Active Allyship Resources

Provides curated learning tools supporting allyship development, inclusive practices, and equity-focused personal growth.

Padlet: https://padlet.com/zenvet23/gcbu_active-allyship/yewow5b9tixhdwf7

HABRI – Human-Animal Bond Research Institute

Conducts and disseminates research on the therapeutic and wellbeing impact of the human–animal bond.

Website: https://habri.org

International Association of Veterinary Social Workers (IAVSW)

Supports the professional development of veterinary social workers, providing resources intersecting veterinary care and human wellbeing.

Website: https://www.iavsw.org

MCVMA – Multicultural Veterinary Medical Association

Promotes diversity, equity, and inclusion through community engagement, advocacy, and professional development.

Website: https://mcvma.org

One Health Initiative

Promotes global collaboration among veterinary, medical, and environmental professionals to improve health outcomes at the human–animal–ecosystem interface.

Website: https://www.onehealthinitiative.com

Pawsibilities Vet Med

Provides mentorship, training, and DEI community resources supporting inclusive veterinary education and practice.

Website: https://www.pawsibilitiesvetmed.com

PrideVMC

Advocates for LGBTQ+ inclusion and belonging within veterinary medicine through education, support networks, and policy development.

Website: https://pridevmc.org

Project Sticker – Stick With Us

A global veterinary wellbeing movement that centers connection, harm reduction, and peer support to reduce suicide risk.

Website: https://stickwithus.vet

Society for Veterinary Medical Ethics (SVME)

Promotes ethical awareness, discourse, and scholarship across veterinary medicine.

Website: http://www.svme.org

The Veterinary Kaleidoscope (TVK)

An Australian organization dedicated to DEI education, community dialogue, and leadership development in veterinary medicine.

Website: https://www.tvk.org.au

Vetquity

Supports diversity and inclusion across veterinary professions in Australia through community engagement and resource development.

Website: https://www.vetquity.com.au

vetThrive

A wellbeing organization providing training, systems support, and resilience-building programs for veterinary teams.

Website: https://www.vetthrive.org

World Small Animal Veterinary Association (WSAVA) – Global Mental Health Resources

Supports veterinarians worldwide with mental health advocacy, education, and international wellbeing campaigns.

Website: https://www.wsava.org

Section 14: Growth, Restoration & One Health Opportunities

Loop Abroad

Offers ethical international travel programs that connect veterinary professionals and students with wildlife, conservation, and One Health initiatives across the globe.

Website: https://loopabroad.com

Pura Vida Veterinary Wellness
Provides restorative veterinary wellness retreats that combine mindfulness, connection, coaching, and personal renewal.
Website: https://www.puravidavetwellness.com

VetVacationCE
Offers destination-based continuing education that pairs high-quality clinical learning with rest, exploration, and personal rejuvenation.
Website: https://vetvacationce.com

Veterinary Leadership Institute – VLE Trek
A transformational, outdoor-based leadership experience designed to build self-awareness, communication skills, resilience, and authentic connection among veterinary professionals.
Website: https://www.veterinaryleadershipinstitute.org/veterinary-leadership-experience

Wild Sun Rescue Center – Veterinary Wellness Retreats (Costa Rica)
A wildlife rescue and rehabilitation center offering hands-on One Health experiences for veterinary students and professionals.
Website: https://wildsunrescue.org

World Vets
A global nonprofit offering international volunteer opportunities, disaster response deployments, and humanitarian veterinary care.
Website: https://worldvets.org

Worldwide Vets – THRIVE Retreat
Provides structured veterinary wellbeing retreats focused on emotional resilience, stress recovery, and personal renewal for veterinary professionals.
Website: https://www.worldwidevets.com/wellbeing-retreats/thrive-wellbeing-retreats-for-vets

Section 15: Coaching Support

Amelia Knight Pinkson, VMD – Life Boost with Amelia
Provides science-based behavior change coaching, emotional wellbeing strategies, and personalized support to help veterinary professionals create healthier, more fulfilling lives. Focuses on sustainable habit formation and self-compassion.
Website: https://lifeboost.today

Corey Blake, DVM – The Institute for Veterinary Wellbeing
Offers leadership training, emotional resilience coaching, and programs designed to improve workplace culture and personal wellbeing in veterinary medicine.
Website: https://theivw.org

Dr. Emily Singler – Parental Leave & Career Coaching
Supports veterinary professionals navigating parental leave, return-to-work transitions, and career development. Offers structured decision-making tools and individualized guidance.
Website: https://www.emilysinglervmd.com/coaching-consulting

Dr. Jen Brandt – Communication & Conflict Transformation
Provides evidence-based communication coaching, conflict-resolution training, and leadership development rooted in psychology and mediation.
Website: https://www.avma.org

Dr. Sarah Wooten – Leadership & Storytelling Coaching
Offers leadership training, communication coaching, and storytelling skill-building to empower veterinarians as effective, confident professionals.
Website: https://drsarahwooten.com

Full Circle Lab – Crystal Stokes
Provides neuroscience-informed leadership coaching, team dynamics training, and transformational development focused on emotional intelligence and self-awareness.
Website: https://www.fullcirclelab.com

Get MotiVETed – Quincy Hawley, DVM & Renee Machel
A well-being coaching organization that helps veterinary professionals build resilience, cultivate joy, and develop sustainable personal wellness practices.
Website: https://getmotiveted.com

Heartstorming Wellness – Sonja A. Olson, DVM
Provides coaching, workshops, and wellbeing education grounded in compassion, neuroscience, and practical strategies for sustainable veterinary careers.
Website: https://www.sonjaolsondvm.com

Jennifer Edwards, DVM – Leadership & Energy Coach
Supports veterinary professionals through intuitive coaching, emotional energy management, and leadership skill building.

Julie Cappel, DVM – The Veterinary Life Coach

Offers mindset coaching, goal-setting support, and personal development guidance for veterinarians seeking confidence, balance, and wellbeing.

Website: https://www.theveterinarylifecoach.com

Kelly Cooper – Reality Vet Coaching & Consulting

Supports veterinary practice owners get out of survival mode and build clear, sustainable practices by strengthening leadership, sharpening business skills, and creating systems that actually work in real life.

Website: https://www.realityvet.com

Melyssa Allen – Healthy Living Doesn't Have to Suck

A wellbeing coach supporting veterinary professionals in building healthier lifestyles through compassionate mindset work and realistic habit design.

Website: https://healthylivingdoesnthavetosuck.com

Natalie Fayman, DVM – No Limits Coaching Now

Provides life and career coaching for veterinarians using mindset strategies that foster confidence, clarity, and empowered decision-making.

Website: https://nolimitscoachingnow.com

Petra Agthe – Vet Mind Works

Offers compassion-focused coaching, emotional regulation tools, and wellbeing strategies tailored to the veterinary profession.

Website: https://www.vetmindworks.co.uk

Reviving Veterinary Medicine – Marie Holowaychuk, DVM

Provides veterinary wellbeing coaching, workshops, and resilience training grounded in evidence-based mental health practices.

Website: https://revivingvetmed.com

Vet Empowered – Dr. Katie Ford & Dr. Claire Grigson

Offers coaching programs centered on identity, confidence, and authenticity to support veterinarians in creating fulfilling, self-aligned careers.

Website: https://vetempowered.com

Section 16: Podcasts & Media

Bent Minded Podcast

A podcast exploring mental health, wellbeing, and personal growth within high-stress professions, including veterinary medicine.

Website: https://www.bentminded.com

Blunt Dissection Podcast – Dr. Dave Nicol

Features long-form interviews with veterinary leaders, innovators, and thinkers, exploring career development, leadership, and personal evolution.

Website: https://drdavenicol.com

Clinician's Brief Podcast

Provides concise, clinically relevant insights and discussions intended to support high-quality veterinary medicine.

Website: https://www.cliniciansbrief.com/podcasts

Dr. Andy Roark – Cone of Shame Podcast

Covers clinical topics, communication strategies, and personal development through engaging interviews and discussions.

Website: https://drandyroark.com/podcast

Hill's The Vet Centric Podcast

Focuses on veterinary practice insights, leadership, and industry trends that support clinical excellence and team wellbeing.

Website: https://www.hillsvet.com/vet-voice/vet-centric-podcast

NAVC VetFolio Podcasts

Provides CE-oriented discussions on clinical care, communication, leadership, and professional wellbeing.

Website: https://www.vetfolio.com/podcasts

The PAWSitive Leadership Podcast

Explores leadership, wellbeing, and positive team culture with practical tools for improving veterinary workplaces.

Website: https://pawsitiveleadershippodcast.com

Vet Med Wellness & Leadership Podcast – Dr. Marie Holowaychuk

Shares wellbeing science, leadership tools, and strategies for veterinary professionals seeking healthier, more resilient careers.

Website: https://podcasts.apple.com/us/podcast/vet-med-wellness-leadership/id1622113993

Vet Vault Podcast

Provides candid conversations with veterinarians around the world about career challenges, wellbeing, and growth.

Website: https://thevetvault.com

Veterinary Innovation Podcast

Features interviews exploring technology, innovation, and the future of veterinary practice.

Website: https://veterinaryinnovationpodcast.com

Veterinary Journeys Podcast

Shares personal stories and career pathways that illuminate the diverse experiences of veterinary professionals.

Website: https://www.veterinaryjourneyspodcast.com

Veterinary Mentorship Podcast

A podcast dedicated to supporting early-career veterinarians through mentorship insights and practical career guidance.

Vet Tech Café Podcast

A show dedicated to veterinary technicians, exploring wellbeing, professional identity, clinical skills, and industry issues.

Website: https://www.vettechcafe.com

VetThrive Podcast

Shares practical tools and wellbeing conversations to support personal and organizational wellness in veterinary medicine.

Website: https://www.vetthrive.org/resources

Section 17: Research References, Articles & Additional Reading

AVMA – Wellbeing & Workplace Studies

Provides research, surveys, and data insights on veterinary wellbeing, burnout, job satisfaction, and workplace culture.

Website: https://www.avma.org/resources-tools/wellbeing

BVA & RCVS – Veterinary Workplaces & Mind Matters

Offer UK-based research and initiatives addressing mental health, suicide prevention, and workplace culture improvement.

Websites: https://www.bva.co.uk/take-action/good-veterinary-workplaces/
https://vetmindmatters.org

CVMA – Health & Wellness Resources

Provides Canadian wellbeing, mental health, and workplace safety resources for veterinary professionals.

Website: https://www.canadianveterinarians.net/veterinary-resources/veterinary-health-and-wellness-resources/

CDC – Worksite Health Promotion

Provides evidence-based frameworks and tools for improving workplace wellbeing initiatives.

Website: https://www.cdc.gov/workplacehealthpromotion

Compassion Fatigue Awareness Project

Offers tools and education to help professionals understand, prevent, and recover from compassion fatigue.

Website: https://www.compassionfatigue.org

HABRI – Human-Animal Bond Research Institute

Advances research demonstrating how animal relationships support human mental and physical wellbeing.

Website: https://habri.org

Merck Veterinary Wellbeing Study

Provides widely referenced data on burnout, mental health, job satisfaction, and wellbeing factors in the veterinary profession.

Website: http://vetwellbeing.com

Mental Health First Aid Research

Offers global mental health education research and frameworks supporting community-based crisis recognition.

Website: https://www.mentalhealthfirstaid.org

NOMV – Veterinary Mental Health Research

Shares research and data collected to better understand mental health trends in the veterinary profession.

Website: https://nomv.org

One Health Initiative

Publishes interdisciplinary research promoting collaborative approaches to human, animal, and environmental health.

Website: https://www.onehealthinitiative.com

Psychological Safety – Amy Edmondson

Provides thought leadership and research on how psychological safety drives team performance and wellbeing.

Website: https://psychsafety.com

SAMHSA – Trauma-Informed Care Framework

Offers research-supported frameworks for implementing trauma-informed practices in clinical and organizational settings.

Website: https://www.samhsa.gov

Secondary Traumatic Stress Consortium

Shares emerging research, best practices, and organizational strategies for mitigating STS in helping professions.

Website: https://www.stsconsortium.com

VetBurnout.com

Provides burnout-prevention research and tools tailored to veterinary medicine.

Website: https://www.veterinaryburnout.com.au

Veterinary Leadership Institute

Conducts leadership research and develops evidence-based programs that support psychological safety and team wellbeing.

Website: https://www.veterinaryleadershipinstitute.org

WSAVA – Global Mental Health Resources

Provides international guidelines, research, and wellbeing initiatives for veterinary teams.

Website: https://wsava.org

Section 18: Recommended Books

Agthe, Petra – The "It's Okay" Coaching Card Deck

A reflective tool designed to help veterinary professionals build emotional awareness, self-compassion, and resilience. Includes prompts that encourage introspection and emotional grounding.

Website: https://www.vetmindworks.co.uk

Allen, Melyssa – Healthy Living Doesn't Have to Suck
Presents practical and compassionate strategies for building sustainable well-being habits without perfectionism or burnout. Focuses on mindset, behavior change, and realistic goal-setting.
Website: https://healthylivingdoesnthavetosuck.com

Andrew, Paul – The Team-Building Book
Offers practical exercises and frameworks to improve team cohesion, communication, and interpersonal trust. Useful for leaders and facilitators.

Bariso, Justin – EQ Applied
A practical guide to understanding emotional intelligence and applying EQ skills in real-life interactions, particularly in high-stress professional environments.

Brown, Brené – Atlas of the Heart
Maps the language of human emotion to help readers understand and articulate their internal experiences with clarity and empathy.

Brown, Brené – Dare to Lead
A research-based approach to courageous leadership, vulnerability, and team trust-building – relevant to veterinary leaders seeking healthier workplace cultures.

Cialdini, Robert – Influence: The Psychology of Persuasion
A foundational text on understanding human behavior, motivation, and decision-making processes.

Cohen, Gregg – Leadership for Veterinary Professionals
Provides accessible leadership principles tailored specifically to veterinary teams, including communication, culture, and decision-making strategies.

Edmondson, Amy – The Fearless Organization
A groundbreaking work on psychological safety – explaining how trust, openness, and learning cultures enable high-performing and resilient teams.

Fleming, Audrey – The Resilient Vet
Focuses on personal resilience strategies that help veterinary professionals manage stress and maintain wellbeing.

Goleman, Daniel – Emotional Intelligence
A seminal book introducing EQ as a key predictor of professional success and healthy interpersonal dynamics.

Goodall, Jane – The Book of Hope: A Survival Guide for Trying Times
It can be hard to feel optimistic. And yet hope has never been more desperately needed.

Jane Goodall, one of the world's most famous naturalists, and Douglas Abrams, the co-author of *The Book of Joy*, explore through intimate and thought-provoking dialogue one of the most sought after and least understood elements of human nature: hope.

Grant, Adam – Think Again
Encourages flexible thinking, curiosity, and self-awareness as essential ingredients for effective leadership and personal growth.

Haidt, Jonathan – The Happiness Hypothesis
Explores psychological principles and ancient wisdom to understand what creates meaning, wellbeing, and emotional fulfillment.

Holowaychuk, Marie – A Compassionate Calling: What it really means to be a veterinarian
An unflinching and deeply personal look in the realities of the veterinary profession.

Kabat-Zinn, Jon – Wherever You Go, There You Are
A foundational introduction to mindfulness and presence for stress reduction and emotional grounding.

Kline, Nancy – Time to Think
Outlines a framework for creating thoughtful, respectful conversations that improve clarity, communication, and problem-solving.

Levine, Peter – Waking the Tiger
Explores trauma physiology and healing, introducing principles that support trauma-sensitive care and emotional regulation.

Nagoski, Amelia & Nagoski, Emily – Burnout: The Secret to Unlocking the Stress Cycle (2022)
Provides an accessible explanation of the physiology of stress and actionable strategies for breaking chronic burnout patterns.

Website: https://www.goodreads.com/en/book/show/42397849-burnout

Neff, Kristin – Self-Compassion
Presents research-based strategies for building self-kindness, resilience, and emotional wellbeing.

Newman, Sandra – The Book of Mistakes
A reflective narrative exploring themes of identity, relationships, and personal growth.

Norfolk, Terry – The Veterinary Formula for Wellbeing
Provides veterinary-specific tools and frameworks to support wellbeing, emotional resilience, and sustainable practice.

Olson, Sonja – Creating Wellbeing & Building Resilience in the Veterinary Profession: A Call to Life
Dr. Sonja Olson's first book uniquely combines shared experiences (personal stories) with academic research into the contributing factors of compassion fatigue and how to counter these. Normalizing the conversation in the profession, it provides a wide array of possible solutions to build resilience and to shape a culture of collaboration and support where caregivers can flourish.

Pinkson, Amelia Knight – Life Boost Principles
A science-informed guide to building healthy habits, emotional resilience, and sustainable wellbeing.
Website: https://lifeboost.today

Rosenberg, Marshall – Nonviolent Communication
A foundational text on compassionate communication, emotional awareness, and conflict resolution.
Website: https://www.cnvc.org

Sinek, Simon – Start With Why
Explores how purpose-driven leadership energizes teams and fosters trust, engagement, and long-term success.
Website: https://simonsinek.com

Stokes, Crystal – Full Circle Leadership
Introduces neuroscience-informed leadership practices that strengthen emotional intelligence, team connection, and organizational culture.
Website: https://www.fullcirclelab.com

Wooten, Sarah – The Vet's Guide to Communication
A veterinary-specific communication guide providing practical tools for navigating difficult conversations and building stronger client relationships.
Website: https://drsarahwooten.com

Section 19: Videos, Webinars & Digital Learning Resources

AVMA – Wellbeing Webinars
Provides a comprehensive library of wellbeing-focused webinars addressing stress management, emotional health, communication, and workplace culture in veterinary settings.
Website: https://www.avma.org/resources-tools/wellbeing

Brené Brown – The Power of Vulnerability (TED Talk)
Explores vulnerability as a pathway to courage, connection, and authentic leadership. Widely applicable to veterinary teams and leaders.
Video: https://www.ted.com/talks/brene_brown_the_power_of_vulnerability

Brené Brown – Listening to Shame (TED Talk)
Examines the role of shame in human behavior and performance, emphasizing compassion, connection, and healthy leadership.
Video: https://www.ted.com/talks/brene_brown_listening_to_shame

Compassion Fatigue Awareness Project – Video Resources
Provides videos that explain the emotional and physiological roots of compassion fatigue and offer strategies for prevention and recovery.
Website: https://www.compassionfatigue.org

Dr. Andy Roark – YouTube Channel
Features educational, humorous, and practical videos on communication, leadership, wellbeing, clinical skills, and practice life.
Channel: https://www.youtube.com/user/DrAndyRoark

Dr. Marie Holowaychuk – Wellbeing Webinars
Offers evidence-based webinars on burnout prevention, compassion fatigue, resilience strategies, and wellbeing science tailored to veterinary audiences.
Website: https://revivingvetmed.com/resources

IHI – Joy in Work Framework
Provides training and recorded sessions on building joyful workplaces through systems improvement, psychological safety, and collaborative learning.
Website: http://ihi.org/joyinwork

NOMV Trauma Webinar Series

A veterinary-care-specific webinar trilogy explaining trauma foundations, trauma responses, and healing strategies.

Part 1: https://youtu.be/oup2_he07kc
Part 2: https://youtu.be/ID0UkBLz7Yg
Part 3: https://youtu.be/3yuRy85YyZ4

Psychological Safety – Amy Edmondson Video Library

Provides interviews, keynote lectures, and educational discussions on team trust, learning culture, and high-performing workplaces.

Search: https://www.youtube.com/results?search_query=Amy+Edmondson+psychological+safety

Stanford Medicine – Compassion in Medicine Course

A digital learning program exploring compassion science, emotional intelligence, and the impact of empathy on patient care.

Website: https://cme.stanford.edu/courses/compassion

Kelly McGonigal – How to Make Stress Your Friend (TED Talk)

Presents stress reframing research that empowers individuals to approach stress with resilience and a growth mindset.

Video: https://www.ted.com/talks/kelly_mcgonigal_how_to_make_stress_your_friend

Susan David – Emotional Courage (TED Talk)

Explores emotional agility and the power of acknowledging, accepting, and responding skillfully to emotions.

Video: https://www.ted.com/talks/susan_david_the_gift_and_power_of_emotional_courage

WSAVA – Wellbeing Webinars

Provides mental health, client communication, and resiliency-focused recorded sessions for international veterinary audiences.

Website: https://wsava.org

Acknowledgments

This chapter – and the broader veterinary wellbeing workbook – was made possible through the generosity, insight, and collaboration of many compassionate individuals and organizations dedicated to advancing wellbeing in the veterinary profession.

Thank you to the veterinary professionals, educators, researchers, and leaders who contributed knowledge, resources, and lived experiences to enrich this collection. Your commitment to fostering psychological safety, emotional intelligence, leadership growth, and sustainable professional fulfillment continues to shape a healthier future for all veterinary teams.

To the organizations and initiatives featured throughout this chapter: your innovation, advocacy, and evidence-based work form the backbone of this resource guide. Your efforts uplift the profession and offer pathways toward resilience, connection, and meaningful change.

Finally, heartfelt gratitude goes to the broader veterinary community – students, technicians, nurses, veterinarians, practice managers, social workers, mental health professionals, and industry partners. Your dedication to animals and to one another is the inspiration behind this work. It is my hope that these resources support you on your journey toward wellbeing, clarity, and sustainable professional satisfaction.

Index

For Product Safety Concerns and Information please contact our EU representative GPSR@taylorandfrancis.com Taylor & Francis Verlag GmbH, Kaufingerstraße 24, 80331 München, Germany